THE
LONDON BRICK
COMPANY

Bill Aldridge

Nostalgia Road Publications

CONTENTS

The **Nostalgia Road** Series ™

is produced under licence by

Nostalgia Road Publications Ltd.

Unit 6, Chancel Place

Shap Road Industrial Estate, Kendal LA9 6NZ

Tel. 01539 738832 - Fax: 01539 730075

designed and published by

Trans-Pennine Publishing Ltd.
PO Box 10, Appleby-in-Westmorland, Cumbria, CA16 6FA
Tel. 017683 51053 Fax. 017683 53558
e-mail: admin@transpenninepublishing.co.uk

and printed by

Kent Valley Colour Printers Ltd.
Kendal, Cumbria. 01539 741344

© Trans-Pennine Publishing Ltd. 1998 & 2003
Photographs: Hanson plc (London Brick Company) or as credited

Front Cover: *London Brick purchased just one ergomatic-cab AEC (although others joined the fleet after the acquisition of Marston Valley Brick), and CBM 242C was used as a test-bed for many of the company's experimental delivery systems.*

Rear Cover Top: *Seen at rest, part of the LBC fleet at the Ridgemont Works in the early 1970s. John E. Hardy*

Rear Cover Bottom: *The new Hanson Brick corporate livery on a London Brick Foden..*

Title Page: *AECs, Albions and an LBC bus outside Stewartby works - a just a small part of what was an impressive fleet.*

Left: *July 1961 saw the arrival of two 6x4 Mammoth Major tippers for site work, including 502 EBM. The Ford Consul just visible to the rear is also a member of the fleet.*

Right: *An early London Brick AEC, EW 9243, outside the Duramin commercial vehicle body works with a load of facing bricks.*

INTRODUCTION

In this, the second edition of this book on the London Brick Company, we have responded to the requests from readers of this series to make available a book that has been out of print for some time. Firms like the London Brick Company remain particularly interesting, where a manufacturer develops their own transport operation! In this case the operation was of specific relevance, for it demonstrated how a manufacturer could also become a large-scale road haulier. In the early days London Brick placed their reliance on publicly offered services, notably the railways, but as the road network opened up, it became possible to supply the customers direct.

By such development firms like London Brick could thus avoid the frequent marshalling, shunting or trans-shipping that a wagon-load of goods would have to undergo, even over short distances. The movement of bricks, by dint of their heavy nature, was a type of traffic that the railways could not always effectively handle. After all, very few building sites had their own railway sidings, so road haulage was bound to win out in the end. Initially few of the early lorries were effective at carrying heavy loads over long distances, but as users and manufacturers developed an understanding of each other's needs, progress quickly followed. Indeed, when you compare delivery techniques of today, with those half a century back, one can not fail to be impressed by the very significant levels of progress.

The way that this type of delivery service has developed in the London Brick Company, actually provides a most striking example of progress. To show how innovation continues in brick transportation right up to the present day, we must really begin by giving a brief history of brick-making and the company's early transport history. In such a way we can then provide transport enthusiasts and modellers with an insight into the intricacies of 'growing' a vehicle fleet that was acquired to cater for the varying needs of the builder and merchant.

To many people the name London Brick is inextricably linked with the Fletton Brick industry and of course AEC lorries. Yet how can a company with such a southern name be linked with a small village now situated in Cambridgeshire? In addition why did the London Brick fleet carry the title PHORPRES so proudly on their headboards? What were Hoffman kilns and why were they so important? Where is Stewartby? How can bricks fire themselves?

All these questions and many others will be answered in the narrative that follows. We will explore the brick-making process, the growth of the London Brick Company and the introduction of delivery vehicles and bring the narrative up to the current period. But, to begin with, we must travel nearly 3,000 miles and go back some 6,000 years.

Bill Aldridge,   Stockport

3

Above: *Horse and rail transport at the LBC's No.2 yard near Peterborough c.late-1800s.*

THE ORIGINS OF LONDON BRICK

Brick-making is one of the oldest crafts in the civilised world dating back at least 6,000 years and starting in the Middle East most notably Egypt and Babylonia. Even though the earliest bricks were sun dried, they were durable enough within the warm climates experienced there, and were certainly used in the building of Ur (Abraham's city). The later use of fire to harden the bricks was one of the first milestones on the road to civilisation. Fired bricks were used in Mesopotamia, possibly around 3000 BC. Their use continued through to Roman times, when great use was made of fired bricks as a primary building material throughout their empire in Western Europe. This craft industry was brought to the British Isles by them, but like many other skills and innovations, the craft was lost when the Romans returned to warmer climes. The Roman brick was also much flatter than modern bricks are, and generally being just 1¹/₄" thick it should perhaps be better described as a tile.

The industry was not properly revived until the 13th-century when Flemish weavers settled in East Anglia. Finding there was no local stone to use for the building of dwellings they made use of their brick-making skills in manufacturing hand-made bricks. By the time of Henry VIII the English had perfected the manufacture of bricks to the extent that the famous palace at Hampton Court was almost entirely built of brick. Of course all those bricks had been hand-made and the business of brick-making remained with individuals and small family groups, suffering the peaks and troughs in sales according to local demand.

The brickworks that these individuals established were spread throughout the British Isles, appearing wherever suitable clay was to be found. Initially clay-pit excavation and the manufacture of bricks was not a particularly capital intensive operation and the industry was always very localised. The bricks were very heavy to transport and as they had no great value the market was restricted to about three miles radius from the production site. These pits and works often operated on a seasonal basis and in direct response to demand from builders. According to a statute of 1477 the clay had to be dug by 1st November and then left outside to weather for up to three months to ensure that it did not shrink during the firing process.

Once dug, regularly turned and dried, the clay would be thrown into shallow pits and soaked with water to break down the lumps. The clay would then be kneaded, either by foot or the use of oxen. When in the correct state the kneaded clay would be moulded into a brick shape and then placed again in the air to dry. When these blocks were dry enough to be handled they would be placed in a clamp or kiln and baked using copious amounts of fuel.

Originally all operations were carried out by hand (or foot), but from the 1850s mechanisation began to replace sweated labour. The use of bricks was however subject to the vagaries of fashion, especially amongst the 'new rich'. Then there was the Brick Tax, which imposed a tax of four shillings (20p) per thousand bricks in 1784.

Whilst this may not sound much today, this figure would actually be about £10 in present day values. So one can imagine the horror with which a further increase in 1794 was viewed, although this tax was eventually repealed in 1850. These two facts alone had quite a detrimental effect on the increasing use of bricks. One enterprising brick-maker in Leicestershire commenced making huge size bricks after 1784 to get round the tax, but the Government sealed this loop-hole with its 1794 increase, which also put a limitation or standard on the size of bricks.

With the many types of clay found across the country the brick manufacturers were able to produce a wide variety of bricks in greatly differing styles. With the gradual mechanisation of the industry some standardisation took place and by the late 19th-century the brick manufacturing business fell into four similar, but significantly differing production methods namely; the Plastic Process, the Semi-Plastic, the Stiff Plastic, and the Semi-Dry. For a works to produce 'plastic' bricks the local clay needs to have a moisture content of 20-30% either occurring naturally or requiring water to be added later during processing.

Semi-plastic bricks have a lower water content while the stiff process bricks use a much harder clay, such as that often found in the coal measures and frequently manufactured by mining companies. The first three processes make use of a mechanically driven pug mill to crush the clay with water added to ensure the correct consistency. The clay is then extruded and wire-cut and the stiff clays pressed to form a brick shape. The manufacture of bricks by the semi-dry method (as practiced by London Brick) will be described later in the text.

During the mid 19th-century a number of engineering companies began to design and actively promote machinery for use in the brick-making industry generally in the form of steam driven grinding, moulding and pressing machinery. In this same period the repeal of the brick tax led to an increase in brick production, and in turn this provided the bricks required for the burgeoning towns spawned by the need to house the workers involved in the industrial revolution.

Much of the machinery used in brick manufacture came from the Lancashire town of Accrington, which was typical of many emerging industrial towns and it also had its own famous brand of red bricks. Although the brick-making business was spread across a wide area of the British Isles it is necessary to consider just one area in particular when we look at the origins of the London Brick Company. From at least the 18th-century and probably earlier there had been a brick-making industry in the Peterborough area manufacturing plastic and possibly semi-plastic bricks.

A significant factor that enabled production to rise in this area was the building of the Great Northern Railway through the town of Peterborough. The quality of the goods delivery service offered by this company, helped the manufacturers promote sales of bricks over a much wider area than had hitherto been the case. Brick express services were offered and in addition to London, such trains were run to the industrial towns of Yorkshire and the Midlands.

Above: *As horse-drawn wheeled vehicles were not particularly suited to working in the heavy clays around the brick yards and clay pits, the brick companies were very quick to seize upon railways for internal transport. A variety of standard and narrow gauges were used with motive power varying from steam to horse-power and hand 'tramming'. Electric locomotives were also tried at an early stage, using overhead collection. A good example of this is seen in this 1920s picture of the Marston Valley works railway. Note the cradles used for conveying the brick trucks. Two of these trucks could be carried in a single cradle, and these were run into the kilns on temporary rails laid at right angles to the track.*

Above: *A picture of the clay workings with the huge shale-planer that was devised in 1926. The clay that it is working (known generically as the Lower Oxford Clay) is the uppermost Jurassic formation in England and had been named as such by the geologist Dean Buckland in 1818. The Oxford Clay bed stretched in a broad band from the Humber to Dorset and as it is not faulted in any way it is thus ideally suited to mechanical excavation (although this fact was not important in the early days). Normally the brick-makers throughout the country used the easily obtainable surface clays, but some of the more enterprising Peterborough brick-makers had begun to use the deeper, Lower Oxford Clay. This lower band was shaley and uniform in depth (at least 100 feet in places) and was much better quality than the higher callow (mixed) clays found on or just below the surface.*

The regular opening of brickworks and their subsequent change of ownership marked the beginning of the Peterborough brick-making business. After 1856 it became much easier to open a Joint Stock Company as a means of raising capital to start new companies, this factor became very important later as the new brickworks required larger amounts of capital to buy the new machinery to mass produce bricks. One of these new companies was started by a local draper, James McCallum Craig, He had bought at auction some farmland belonging to Fletton Lodge in 1877 and brick-making commenced there probably in the closing months of 1878. In the following year Craig leased the land to Messrs Searle & O'Connor and in turn a Grantham-based brick-making company called Hempstead Brothers took over the operation of the brickworks. It is fairly certain that Searle & O'Connor (and possibly the Hempstead's) had house building interests in the London area for which a plentiful supply of quality bricks would be necessary.

Soon afterwards the Hempstead's leased a further 30 acres of land, but these new leases banned Mr Craig from being involved in those selfsame works. It would appear from reading between the lines that Mr Craig had made a very important discovery but the Hempstead Brothers wanted to exploit this potential themselves. What had happened was that someone within these works had discovered (whether by trial or error is not recorded) that the local clay had some very special properties!

The brick-makers found that the uppermost layers of weathered callow clay produced soft yellow bricks that had little strength, whilst the lowest layers of clay had too high a lime content which could damage the kilns. However, the depth and uniformity of the middle band of Lower Oxford clay was not its only useful characteristic.

During the processing operation this middle band of clay did not need any additional water to help in the grinding process to break it down. This shaley thick band of Lower Oxford clay changed quite easily on grinding into a granular form and could then be pressed into a block shaped like a brick and was immediately ready for firing. In this state the pressed clay blocks were known as 'green bricks'. The low plasticity of the clay enabled the green bricks to be handled and stacked immediately after pressing. There was no need for a 'pug mill' to grind the clay to a homogeneous consistency by adding water as required in the plastic and semi-plastic brick-making processes and this factor really simplified the manufacture.

However, the vitally important third discovery was made following further experimentation. This discovery confirmed that the natural carbonaceous content of the lower clay meant that, once warmed, the bricks would burn themselves without the need for an external heat source. With a natural moisture content of around 18% and a low plasticity the pressed green brick could immediately be placed in a kiln to be dried by warming before being fired.

In comparison the green bricks made from other clays and by other manufacturing methods needed an extensive drying off period in the open air before being set in the kiln.

This of course involved double handling of the bricks and a more expensive firing operation and, since the production method was slower, more capital was tied up in the operation!

At this point the meaning of the expression 'bricks that fire themselves' becomes obvious since the green bricks did exactly that, as opposed to all other brick manufacturing processes where the bricks need firing by means of a fire. It is certain that not every detail of the manufacturing process noted above was discovered on day one of Fletton brick production. Rather, only a long series of extensive trials led to the eventual perfection of the production process.

The manufacture of bricks made from the Lower Oxford Clay became known as the SEMI-DRY process since no added water was required. Less processing meant less machinery therefore lower costs, faster production and usually more profits for the owners. With this new method of brick manufacture having been discovered and put into production within the confines of the Soke of Peterborough, it was only natural that a local name should be chosen for this style of bricks. Consequently they became known as 'Flettons' after the local village where the first experiments were carried out.

The early history of the present London Brick Company is rather convoluted to say the least and the following notes are more of a shortened summary than a complete history. During the 1880s the Hempstead brothers who were the instigators of the semi-dry brick-making process formed a company called the North London Freehold Land and House Co. This company owned houses and land in London, from which we can ascertain that it was almost certainly involved in speculative house building. Unfortunately the company went into liquidation in 1883, the liquidator being one John Brown.

He tried to sell all the Hempstead's land and brickworks, but was not successful, probably due to a trade downturn at that particular time in the house building market. Eventually a Charles Ormston Eaton bought most of the land and works

He in turn formed The Fletton Brick Company, which much later became Eastwoods Ltd., a company whose name will subsequently reappear in this narrative. One other yard originally auctioned in 1877 became known as Hardy's Yard, although this in turn went bankrupt in 1885, again due to the poor trade in the building industry. One John Cathless Hill bought this yard much later in 1888. This gentleman was again a speculative London house-builder and was attracted to Peterborough by an auction of a half million bricks at a time when there was an apparent shortage of bricks within the London area.

Having bought the brickworks he then formed the first London Brick Company. Unfortunately this company also ran into financial difficulties (like many other building companies and merchants then and now). Luckily the company was allowed to carry on trading and Hill was discharged from bankruptcy in 1915. However Hill, who was ably assisted by his manager Adam Adams, had continued to expand the London Brick Company into the largest company in the industry. After J. C. Hill died in 1915, his son (John Edgar) continued to expand the business.

At this stage of the story we have to look slightly further afield into Buckinghamshire where a Mr. Itter purchased land for a new brickworks. He had originally operated the Kings Dyke brickworks near Peterborough but he soon realised that the final main line railway to be built into London, the Great Central would need bricks for stations and tunnels and other engineering work. Recognising that the proposed line would pass through the Oxford Clay belt, he built his new works at Calvert, with the expressed aim of serving both the railway's needs and the requirements of the speculative builders in West London.

He had guessed correctly, for the railway did indeed consume an inordinate amount of bricks during its construction period, requiring both common and engineering bricks in large numbers. Unfortunately, and rather perversely it was the London Brick Company that received the main order from the Great Central for the supply of no less than 25 million bricks. It is possible, that Itter's Calvert works may have supplied the railway at a later date.

In Bedfordshire, a 'plastic' brickworks had been opened in 1894 on the Midland Railway at Westoning. These were owned by B. J. Harfield Forder, from Buriton in Hampshire, who had started in business after he had inherited a lime-burning business. His success was such that his operation had been extended into the Oxford Clay lands by 1897, with brickworks being opened at Elstow and Wootton Pillinge. All these three sites were highly mechanised to ensure maximum return on capital.

In order to strengthen the management and finances of his firm, Forder looked to the Fletton brick-makers of the Peterborough area and persuaded the Keeble Brothers, George and Arthur, to join him. They in turn invited the then prospective Liberal Parliamentary candidate for Peterborough, a gentleman called Halley Stewart, to join them. Stewart was able to supply much of the required capital since he had recently sold the family oil and seed crushing business in Kent (Stewart Brothers & Spencer of Rochester). Unfortunately the Keeble brothers turned out to be speculators looking for short term gains whilst Stewart was far more interested in the long term view.

The Keeble's themselves owned a few small brickworks and these were later taken over by Mr. Stewart under the Forders banner. When Forders became a limited company in 1900, Stewart was appointed chairman; he brought other members of his family into key positions and continued building up the company so that by 1910 it was producing and selling 48 million bricks per year. After World War I, the Fletton Brick industry (as the semi-dry brickworks in the Peterborough, Bedfordshire and Buckinghamshire areas had come to be known) operated in four main groupings. These were the London Brick Co., The Forders group, the Itter's companies and a looser group headed by the United and Northam Brick companies under the control of J. W. Rowe who was friendly with the Stewart family. Most of the smaller brickworks had, by this time, been taken-over by one or another of these four companies during the previous 10-15 years.

The cyclical nature of the business with its boom and bust plagues still unsettled the whole industry, and the Fletton process suffered from rather high fixed costs due to the need to keep kilns in operation 365 days a year. To try to bring a stable pricing policy to the brick manufacturing business there was a suggestion that the London and Forders brick companies should amalgamate, and this eventually happened in 1923.

Shortly afterwards the United group joined the combine, following the death of one of their leading lights. The new company traded as the London Brick Company and Forders Ltd., with Percy Malcolm Stewart as chairman. As such they were able to place a stabilising hand on the industry as regards output, pricing policy and better use of clay reserves.

The new company also bought a controlling interest in Itter's in 1928 although it was not until 1937 that Itter's ceased to exist as a separate entity. In 1936 the amalgamated firms commenced trading simply as The London Brick Company Ltd. and in due course they changed the name of the works at Wootton Pillinge to Stewartby in respect of the leadership given by the Stewart family to the industry over a long period. In the 1930s the LBC was well ahead in social thinking by offering better conditions and higher pay levels for their staff in accord with their philanthropic ideals (rather like the Quakers and the Cadbury family).

Top Left: *The use of the railways required the London Brick Company to have its own vehicle fleet at an early stage, but these were railway trucks known as private-owner wagons. Two types, both built by Charles Roberts & Co, are seen here in the 1940s, but both show different liveries.*

Centre Left: *This view shows one of the LBC's main competitors, the Marston Valley Works with an AEC and a Commer waiting to depart in the 1960s. The dispatch office seen in the centre of the picture was a fascinating, busy place. In charge was Mr. Collins, who (with his staff) was responsible for allocating the runs according to orders for the two shifts at six and seven o'clock. The 6am drivers were all on Bedford lorries, the big Bedford's (the 7-ton S-Types) carrying 3,500 bricks, the smaller 5-ton O-Types managing 2,500. Their main journeys were to the London area or the more local runs. This enabled them to return to the works for a second run. The 7am shift utilised the rest of the fleet; small Bedford's, the four-wheeled AECs and ERFs and all the eight-wheelers, which were used for transporting the orders to the more distant sites!*

Bottom Left: *The village of Stewartby was built by the company around the works in the 'Garden Village' style of Letchworth or Welwyn Garden City with houses rented by the workers. A number of houses were also erected in the Peterborough area. Sir Malcolm Stewart remained in control until 1950 when the chairmanship passed to Mr. (later Sir) Arthur Worboys and then in 1966 to another Stewart, Sir Ronald Stewart.*

PROGRESS

Between 1925 and 1950 the London Brick Company was very ably guided by Sir Malcolm Stewart. In that time the company's size and production levels kept growing, hampered only by the shortage of materials and manpower during World War II. In fact the conflict had mixed effects on the company, initially they enjoyed large brick orders in connection with the airfield construction programme.

However, with the country's attention turned to other things (like preventing a German invasion) the demand for bricks slackened off in early-1940. Unfortunately, when sales finally took off again, there was little transport available to move the bricks. Accordingly brick-making continued at low levels for the rest of the war, and by 1945 the company had built up a stock of 500 million bricks; though these were soon sold.

The country then entered its own boom and bust period with the government attempting to control the economy and defend the pound. This first post war recession had its roots in the Marshall Aid plan, whereby the Americans were to help the war-torn European economies recover from the ravages of war.

Unfortunately after the Republicans won the 1946 Congressional elections, American price controls were revoked and prices rose and this eventually led to a run on the pound bringing with it an austerity programme that dogged both Britain's first post-war Labour government and the building industry alike.

The British economy soon bounced back, only to fall prey to another recession in 1949. Devaluation against the dollar followed, but by 1951 the Government was able to suspend Marshall Aid, only to be hit by inflation caused by the Korean War.

It is fascinating to look back even over this short period and discover that events happening half way across the world could have such a serious effect on basic industries like brick-making and transport. Since then Governments of all persuasions have made use of budgets to similar effect, with the building industry utilised simply as a tap to be turned on and off as required by economic circumstances. However, to look on the brighter side, the post war era did herald a new era in brick-making: with better-funded research into bricks, brick-making and customers' requirements.

As related earlier, to some extent it was pure chance that brick making grew to such a large extent in Peterborough, but sadly the history of the Fletton industry is too great to recall here.

Left: *For those who want to know more, the detailed history of London Brick's development is told in Richard Hillier's excellent book* The Clay That Burns, *which was published by the London Brick Company in 1981. The author found this a most useful work of reference, but he would also like to thank members of staff at the company for the background information that was also used. Here one of the company's Leyland Cubs, Fleet No. B6 (EW 9240), is seen prior to delivery.*

One major factor made the growth of the Peterborough brick industry almost inevitable. This was the building of the railway network. The Great Northern Railway arrived from Boston in 1848, and was extended to London in 1850 and then to Retford in 1852. In doing so it put the town on what would become the East Coast Main Line. Another route running north-east gave a connection to the Midland & Great Northern Railway at Sutton Bridge in Lincolnshire, after it commenced operations in 1866.

Inside a very short period Peterborough found itself connected to five separate railway companies, the GNR, the LNWR, the M&GNR, the GER and the Midland. Naturally, these railway links had a significant effect on the growth of the town with the population increasing from 4,000 in 1801 to 66,000 in 1962. We might contrast Peterborough's growth with a similar local town, Stamford. This had no main line connection, merely a lowly branch line, and had a population growth of just 3,000 people in the corresponding period.

In addition to passengers, the railways also carried goods. Indeed the provision of goods transport was the railway's *raison d'etre*, and for quite a long time this was the only means of moving large quantities of goods quickly over any distance inland. With the direct connection to London, the capital city quickly established itself as the major market for Peterborough-produced bricks.

As production increased during the 1930s following various take-overs and mergers the Fletton brick came to be sold over a much larger area. By now the Fletton industry was able to compete with other brick types and manufacturers across the country rather than being involved in competing with itself.

Left: *From time to time the kilns at the various works would have to be repaired or replaced, and sometimes large-scale building works had to be undertaken on the site. Few pictures survive to record this vital, but routinely mundane type of operation, so the picture seen here is a real gem from the London Brick Company archives. The vehicle seen tipping sand at the new kilns is a Dennis Pax in the employ of E. H. Lee (Hauliers) of New Road, Woodstone, Peterborough. An interesting aside is that even back then this company had at least two telephone lines (Peterborough 4853/4). The firm is still operating in the area today, and although the four-digit 'phone number has long gone it still forms part of this company's current number. The lorry has fleet number 39 and is painted with a maroon cab and red body; the lettering was yellow with green relief. The original Mr. Lee moved from Merseyside to work in the brickworks, and later began transporting bricks for the LBC. As the firm grew in size they started transporting waste bricks from the works, taking these to local farms where they were used to make better farm roads.*

Above: *This AEC Matador was from one of the first batch of post-war AECs purchased by London Brick. With fleet No. K23, EBM 115 had a 7.7 litre engine and was designed to haul a trailer.*

MECHANISATION OF THE INDUSTRY

Despite the basic nature of the brick-making process not having changed over 6,000 years, the introduction of machinery to the industry resulted in the increased production of better quality bricks at a lower cost to the consumer. To start at the beginning of the manufacturing process we need to return to the clay pit where initially all the clay was cut by hand, with the only machines being in the form of rail wagons to move the clay to the works.

Within the pit there was an 'uncallowing' gang who had the responsibility to remove the overburden. The 'getters' (who levered the clay off the working face) stood and worked at different levels to ensure that a good mix of clay went forward to the grinding operation. The clay pits came to be known as Knott Holes after the hardness of the 'Knotts' of clay. In the early days the workmen cut down through the overlying clay until a change in consistency checked him just as a knot in wood stops a saw.

It was perhaps a foretaste of things to come that the introduction of machinery into the clay pits themselves was occasioned by a strike! In 1902 an eight-week long withdrawal of labour by the 'getters' forced the London Brick Company (then under the leadership of Mr. J. C. Hill) to purchase a Ruston & Proctor steam navvy to replace some of the men. Although this particular machine was somewhat restricted in its ability to replace the men totally, it proved that the principles of mechanised clay extraction were both sound and worthy of further examination.

In certain pits steam-driven face shovels were purchased where the working face was not too high. In other pits where the clay bed was deep, a 'shale planer' was used (this type of machine was fitted with an almost vertical jib to shave the clay off the working face and is seen on page 4). Removing the overburden became a job for large draglines; after the 1920s some of these were diesel powered, whilst many others were operated by electricity as they are today. Where the clay pit lay close to the brickworks then horses or ponies might be used to drag the wooden waggons.

In other pits steam engines drove continuous haulage chains to drag flange-wheeled waggons to the works. These chains and waggons were later replaced by belt conveyors or aerial ropeways, which were ideal for transport over both long distances and difficult terrain. This type of system was also of great value where several roads had to be crossed. Within the brickworks themselves the machinery was originally driven by steam engines similar to those installed in cotton mills. Power transmission was either by shafting and belts or direct by rope from the engine flywheel, later these same engines could be used to drive generators with electric motors driving individual machines. The steam engines in turn were eventually replaced, sometimes by gas engines, but more likely by mains electricity.

The first machine in the Fletton brick-making process was the 'kibbler'. This reduced the lumps of clay or knotts to a uniform size, before they entered the grinder or hammer mill. The ground clay then passed through electrically heated screens and into the presses. There were two separate operations here. Initially the ground clay was pressed into a brick shape in a mould in which two plates slid up and down, it was then given a second press in a similar fashion to compact the clay even more. This operation of press and repress was completed in two cycles of the machine and became known colloquially as 'four press' machines, hence the slightly classical 'PHORPRES' trade name applied to bricks produced by the London Brick Company. The common Fletton brick then passed directly to the kiln, but those pressed bricks destined to become facing bricks underwent a further process. To alter the texture of the brick face, the bricks passed under a series of rollers to create a pattern on the brick before various pigments and sand mixtures were literally blasted onto the face and both headers of the brick. This gave a wide variety of styles and colours that were progressively introduced after the very first facing brick was produced in 1922. That first facing brick was called the Rustic and, remarkably, it is still in production today.

The original kilns used in the Peterborough area were of the intermittent type, which could fire all the variety of bricks that were manufactured locally. Unfortunately these fairly simple kilns were not really suited to firing semi-dry bricks made from the local clay because of the emission of large quantities of fumes and smoke. Whilst the smoke might have been acceptable by the standards of the time, the oily, carbonaceous fumes were frowned on by many of the local residents. In fact an injunction was taken out against the Hempstead company as early as 1881 preventing their brickworks from emitting 'noxious fumes'.

Luckily, by this time there was an alternative type of kiln, which was based on a regenerative principle whereby waste heat was reused and the noxious content of the gas was gradually reduced, the final emissions were discharged through a higher chimney and these were deemed more acceptable. However, the use of intermittent kilns continued around Peterborough for some time, although these may well have been situated further away from the population centres.

In 1856 Friedrich Hoffman had invented the continuously burning kiln and after it was patented in 1858 it became one of the most important innovations in the whole history of large-scale brick-making. The Nottingham Patent Brick Company initially obtained the exclusive rights to the Hoffman Kilns and they had six of them in operation by the 1860s. This new process made use of a number of chambers within a circular kiln with a central chimney. As noted above, the waste heat and gases were used to both pre-heat the air that was to be used for combustion and to warm the green bricks that were drying. The injunction against the Hempstead brothers forced them to invest in the construction of a modified version of the kiln known as the traverse arch Hoffman kiln. The operation was such that there was only one fire and this was moved round the kiln from chamber to chamber, rather than the bricks moving through the kiln.

In front of the fire the new green bricks were being gradually warmed, whilst in the chambers behind the fire the fired bricks were slowly cooling. This version of the new kiln gave many advantages over traditional methods, and led to further modifications around the turn of the century when the internal flue pattern was modified to allow even greater use of the waste heat and the diminution of noxious gas emission. Production increased considerably! Furthermore fewer 'seconds' were turned out thanks to the more even firing of bricks within individual kiln chambers. Previously these seconds had to be sold at a reduced price or discarded, with more firings needed to obtain sufficient 'perfect' bricks.

A further useful effect of these changes was to reduce the volume and weight of coal used for firing. These continuous kilns were a very interesting concept and a fuller description of their working would not be amiss. The essential principle was that the fire travelled to the bricks and not the other way round, e.g. when the product travels through the kilns. Most continuous kilns contained either 16 or 32 chambers with the latter having two independent circulating fires. When the green bricks were stacked or 'set' in the chambers they contained about 19% moisture and this moisture had to be carefully driven off before the firing/burning process could commence. In the warm chamber the bricks stood for one day with the temperature rising to 60°C. A four or five day drying process then began by introducing warm air from the cooling part of the kiln.

Once dry, the bricks reached the maximum firing temperature of about 1010°C simply with the carbonaceous content of the clay combusting by itself, having been ignited by the main fire gases entering from the previous chamber. It must be emphasised again, that the bricks burned by themselves, and they were not fired in the conventional sense. They were then allowed to cool or 'ease' slightly to 900°C by permitting cool air to enter the chamber. They then had to 'soak' at that temperature for 30-40 hours, before 'Smudge' (poor, low-grade coal) was added to maintain the temperature.

After this period the bricks were allowed to cool and in doing so they acted as a pre-heating unit for the combustion air flowing to the main fire. Subsidiary flues took all the hot air from the chambers that were cooling over the top of the chambers to the bricks that were drying. After firing and cooling, the temporary 'wicket' (which sealed the chamber entrance) was removed to enable the combustion air to enter the kiln. At this point the bricks were ready to be 'drawn ' from the kiln.

All the firing operations were constantly monitored and manual control of the dampers was necessary, as this controlled the speed of movement of the fire from chamber to chamber as well as the rate of air supply to the drying chambers. The water vapour and gases formed during drying and burning passed along flues under the chamber floor to the chimney. The whole process took about eleven and a half days to dry, fire and cool the bricks. The in-built ability of the clay to fire itself helped reduce the production costs, and thus produced a cheaper brick when compared to those manufactured from other clays.

Above: *This AEC Monarch MkII, fleet No. D6, is pictured when new in 1936, just prior to its departure from the Southall works of AEC.*

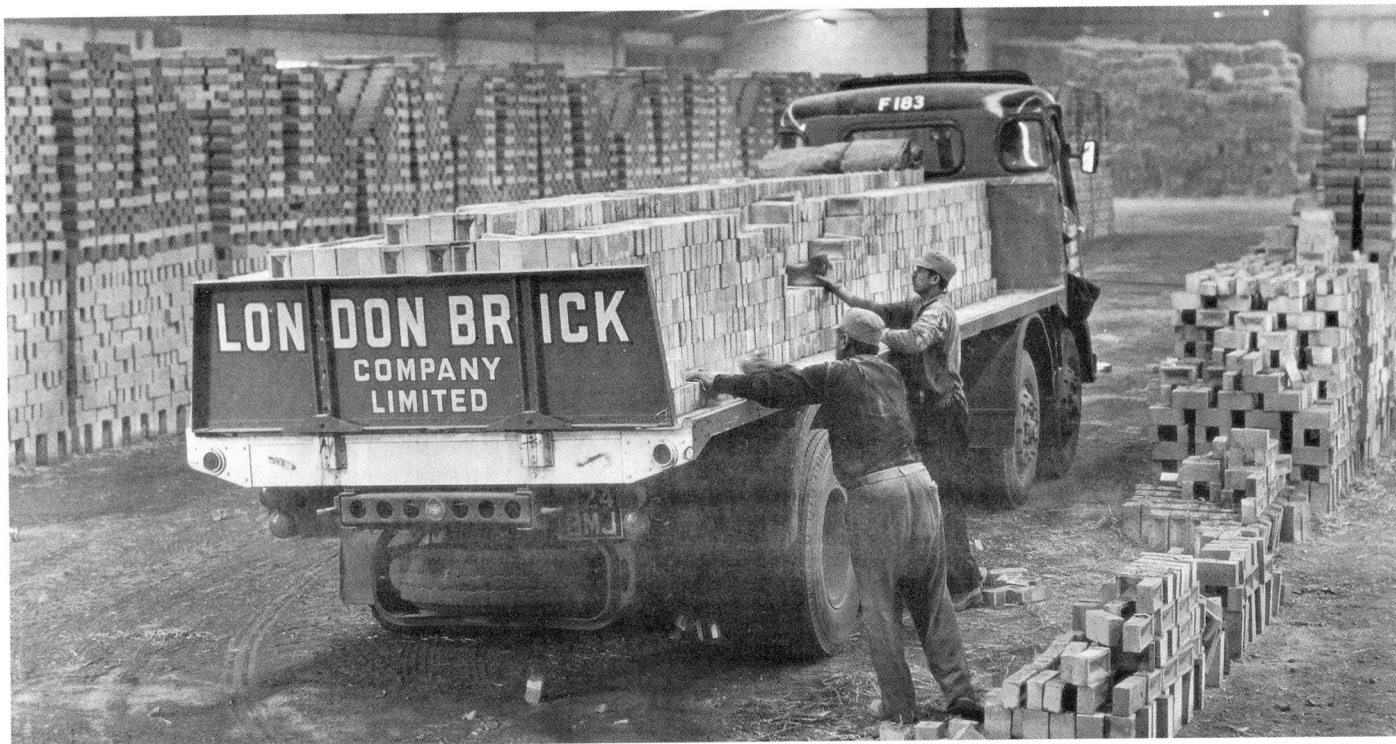

EARLY MATERIALS HANDLING

Up to this point the brick manufacturing process method had been mechanised, but to actually stack the bricks ready for movement from the presses to the kilns the owners relied on a number of men and boys who would transport the green bricks to the kilns. They used wooden barrows with a single wheel known as 'Jollys', on which 90 green bricks could be carried in one journey. Occasionally this function might be transferred to a miniature or narrow gauge railway, but it still required much hand-balling of the green bricks into the kiln.

When the new Marston Valley works were opened they used mobile presses moving from kiln chamber to kiln chamber to reduce this double handling. Much later small petrol or diesel trucks and trailers came into use to move the green bricks. Yet, whichever method was used there was still the need to remove the unfired bricks manually from the presses, transport them to the kiln and then re-stack them in the chamber in such a way as to allow the air and heat to circulate freely to fire the bricks. Having been fired the bricks would have to be manually removed from the kilns and stacked for movement to the stockpiles, then later moved again as they were loaded onto road or rail vehicles for delivery. The eventual solution to the multiple handling problem was quite brilliant.

The genesis of the plan came from two London Brick directors, James P. Bristow and Col. W. D. Rowe, who in 1947 had realised the potential of the forklift truck. The company imported a Hyster truck from America and replaced the two forks with a series of tines. These tines were fitted with longitudinal inflatable bags and the system relied on the green bricks being stacked as they came off the presses in the same fashion as they would when they were being set in the kiln with air spaces between the bricks.

By doing this and placing the lowest layer of bricks vertically according to a template, the tines could run between the bricks and when the bags were inflated they would tightly grab the bottom layer and lift the whole load straight into the kiln. The lower, vertical bricks were commonly known as soldiers. Of course the kiln entrances had to be widened, but this was a small price to pay for the huge increase in productivity.

To give an example of the time saved, in the 1940s loading a chamber (45,000 bricks) by hand used to take up to one and a half shifts to complete. With fork-trucks it took two and a half hours London Brick's first kiln to be converted was at the Norman Cross works near Peterborough and the historic day was 8th August 1947 when the bricks were drawn from Kiln 3 chamber 9, for the first time ever by mechanised means.

Left: *A considerable degree of manual labour was required to load the vehicles, and even after forklift trucks had been introduced in the kiln work, loading of lorries such as Mammoth Major 524 BMJ (fleet no. F183) was done by hand. Note the sloping stacks of brick, fashioned to fit the kilns.*

Top Right: *Interior view of the transport department garage at the Fletton Works in the late-1930s not long after its opening. For a time the entire fleet was garaged under cover, but as the number of vehicles grew the majority had to stay outside overnight whilst the covered accommodation was reserved for those requiring attention.*

Centre Right: *A Shelvoke & Drewry forklift truck moving bricks at the works.*

Bottom Right: *Loading bricks on to D123 (AEW 752) at the kilns, with a forklift and dumper in attendance. Note the other works in the distance.*

The green bricks were 'set' and moved in two distinct patterns, one of 1,000 bricks making the lower part of a stack in the kiln chamber and the second with 900 bricks stacked and shaped to fit in the chamber. After firing the bricks had lost a lot of weight and the 1,900 bricks could be 'drawn' by one forklift truck. In passing it is interesting to note that Mr. Bristow was the grandson of the gentleman who first complained about the smell from the chimneys in Peterborough in 1881.

These fork trucks themselves deserve to be looked at in detail since the production models came from a manufacturer who was more used to making dustcarts. Whilst the prototype had been imported, the Government would not allow any more to come from America and LBC had to look elsewhere to place the subsequent orders. The manufacturer chosen to take over from the Americans was Shelvoke & Drewry who had great experience with hydraulics and were based in Letchworth relatively close to the works. This company's famous designer C. K. Edwards came up with the Freightlifter design, which was based more on heavy duty truck engineering rather than the lightweight fork lift trucks to be found slowly gathering popularity in factories at the time.

Capacities of the Freightlifter varied from 12,000lbs (5.5 tonnes) to 18,000lbs. (8.2 tonnes) and, because of their versatility, the LBC bought a total of 170, some of which remained in service for 21 years. In 1962 the fork truck range was totally redesigned and called Defiants, but these had little in common with the road vehicles that S&D were producing at the same time.

The range was later sold to Rubery Owen, who made Conveyancer fork trucks, before it then passed to Coventry Climax who later traded as Kalmar. The introduction of fork lift trucks to the manufacturing process rapidly spread to all aspects of handling, loading (both road and rail vehicles) and also for stockpiling quickly and neatly when necessary. This stockpiling was very important to cover the cyclical nature of the building industry, since it was very expensive and difficult to restart kilns once they had been allowed to go out.

CLOUDS ON THE HORIZON

Right up to the late-1960s the Fletton brick had been the mainstay of many tens of thousands of housing projects, the Fletton common brick formed the internal walls whilst the Fletton facing bricks were used on the external walls. Despite the relative cheapness of the Fletton brick it was not universally used throughout the country due in part to its lack of frost resistance in certain prevailing weather conditions. The London Brick Company also made a virtue out of the fact that there was a slight difference in colour and texture between bricks that were fired within one chamber and also differences between bricks coming from different chambers within the same Hoffman kiln.

The company made great play of the fact that the bricklayer could form patterns in walls by carefully picking and choosing the bricks he was about to use. Overall the company had the ability to meet huge orders very quickly and some estimates show that up to 85% of southern England's housing stock has been built with Fletton bricks over the last 100 years. However, the very cheapness and consistency of the brick did not overcome the need for even cheaper forms of housing that were required during the 1950s. The Government's answer to the slum clearance plans and the corresponding housing shortage was to initiate the building of tower blocks.

Above: *An early AEC Mercury is pictured delivering bricks to Shell-Mex House in London. A careful examination of the picture would give modern-day 'Health & Safety' officers a 'field day'!*

By their very nature these tower blocks needed to be system-built using pre-formed concrete panels for which no bricks were required. Not wishing to be outpaced the company rapidly introduced the brick faced panel, which was designed to be more appealing to the eye than the aggregate rendered panels that appeared on many of the earliest tower blocks. Whilst the theory was all right, the actual manufacture and especially the transport and manoeuvring of the blocks into position without causing damage to the panels was fraught with problems; consequently the panel business did not thrive long term.

As the 1950s gave way to the more affluent 1960s there was a definite trend towards more individualism and away from the regimented tower blocks and housing estates. The private house-owner generally wanted a dwelling that looked a little up-market and unfortunately, in many cases, the Fletton brick was not perceived as the brick for the job. As the 1970s approached there was a move towards the use of plasterboard internal walls in houses and offices and the use of breeze blocks for internal walls. With facing bricks thereafter being supplied simply to make a house look nice, the sales of the common Fletton brick declined.

However, this was somewhat counterbalanced by an increased demand for a range of different facing bricks. Additionally the architects were very keen to ensure that the bricks that they ordered were consistent and did not vary from one batch to another. At the same time the company could not afford to produce bricks that they could not sell due to variations in colour.

This led to an increasing reliance on the research department to find new ways of producing a wider variety of bricks, in smaller batches, yet with total consistency between batches. This was quite a tall order for what was essentially a cheap product. The ideas and plans that the research department came up with impinged on all components of the brick manufacturing process and many changes came to be wrought in the never-ending process of chasing the moving target of customer preferences.

The research people made many recommendations at this time, but in a nutshell their main finding was that the manufacturing had to be updated. Better handling of bricks was also required, both at the works and on the building site, and by implication during transit. It was also clear that despite the skill of the kiln operative, the firing of kilns needed to be much better controlled.

Even the best run Hoffman kilns were found to have great variations in the quality of bricks between different chambers and different weather conditions. Yet just minute variations in temperature could cause problems and lead to the rejection of many bricks. One of the more surprising comments from the research team was that the largest Hoffman kilns did not necessarily produce the best bricks and an optimum size of kiln was designated.

Up until the 1960s the company had retained and kept open all of the existing works both large and small to ensure that the market share could be retained and large orders quickly executed. However, by this time the company began to realise that some of these works were definitely outmoded and expensive to run, but they still kept them in operation.

Needless to say the market was still volatile and forward planning was never easy, especially when the Government continued to use the building industry as a regulator for the economy. To give the company their due they did realise the effect that the higher production costs at the older works were having on profits. Accordingly there was a definite movement towards larger production units, along with the closure of a number of outdated works. The 'modernisation' was achieved through the rebuilding of the larger works especially during the 1970s and 80s, with a degree of centralisation in both manufacture and transportation being achieved as a result.

With the introduction of the newer works many improvements came about; for instance certain additives were introduced to the clay to help the consistency of the bricks whilst being fired. New, modern design kilns were built giving much greater control over the firing, new methods of handling bricks reduced damage, and transportation issues were considered as an integral part of the 'quality' control. But, more importantly the company was actually taking steps to move the common Fletton brick up-market in order to meet the needs of the industry!

Although this narrative has so far concentrated on the production of bricks there are two other clay-based products that the company sold over a long period: namely pipes and blocks. The clay pipes had a long and illustrious history in the draining of agricultural land to make the fields suitable for crop production. Unfortunately the laying of the pipes was a very labour intensive job and the introduction of the continuous plastic pipe, which could simply be unreeled from a drum and laid in the ground put paid to any long term future for the clay pipes. The blocks came in many shapes and sizes, their main use being for flooring and partitioning in high rise buildings, especially in the London area.

These blocks could also be made up into bespoke flooring beams at the works prior to delivery, but given the quality of industrial building at the time the beams would often arrive on site only for the contractor to find that they would not fit! With the additional problem of the blocks being very susceptible to damage it was decided to forsake that particular market. The blocks and pipes were manufactured mainly at Arlesey, Beebys and Warboys works.

Left: *To facilitate better loading conditions, and move the loaders in out of the elements, a new shed was constructed. In it we see Fleet No. G20, an Albion Chieftain (HMJ 260) being readied for its next trip. This particular Glasgow-built lorry lasted from 1950 to October 1963.*

PROGRESS THROUGH ACQUISITION

By the 1960s only four principal manufacturers were to be found in the Fletton brick industry, these were the LBC, Eastwoods, Marston Valley Brick Company and the Whittlesey Central Brick Co. now trading as National Coal Board Ancillaries (Whittlesey) Ltd. (Please note that the spelling of Whittlesey regularly varied between 'sey and 'sea). The Eastwoods company was an old established firm operating initially as builders' merchants who later diversified into brick-making, cement manufacture and concrete tile production. They had taken over Flettons Ltd. in the 1930s and had quickly installed new equipment in the works.

In 1962 Eastwoods came to be taken over by the Rugby Portland Cement Company, who bought a mixed bag of brickworks, gravel pits, cement works and pipe works as well as the builders' merchants operation. In turn the brickworks were sold to the Redland Group who had wide-ranging interests in bricks and tiles. By the early 1970s the clay reserves of Redland (nee Eastwoods) at their works at Kempston, Yaxley and Orton were beginning to run low and rather than invest in new reserves Redland sold out their brick-making interests to LBC in 1971.

Throughout the history of the London Brick Company the management had always been on the lookout for competitors that were vulnerable to being taken over. Amongst many other take-overs the names of the Bedford Brick Company and the Grovebury Brick Company have been noted.

Above: The biggest Fletton brick manufacturing competitor to the LBC in the inter war period was the Marston Valley Brick Co., who were situated at Lidlington in Bedfordshire. This picture of three Bedford TK tippers at work is especially pleasing, and we are actively seeking other Marston Valley (and other brick company) photographs to produce similar books to this one.

The Marston company was started in 1929 by an ex-manager of the LBC, who built a brand new works incorporating all of the most up-to-date equipment. At the time the techniques employed at Lidlington were well in advance of those employed by the LBC group and included brick presses that could move on rail lines to the entrance to each kiln chamber, which considerably reduced the handling of the green bricks. Marston Valley then built a second works, not far away at Ridgmont, which made a very good strong Fletton brick.

Although a reasonably successful firm, they were finally taken over by the LBC in the early 1970s. London Brick had made an offer for the Marston shares in 1968, but had to wait three years for full ownership. Marston Valley had never really been big enough to seriously challenge LBC and they found to their cost that the brick manufacturing techniques of the 1930s, when labour was cheap, were no answer to the wholesale modernisation that was required to retain market share and profitability in the 1960s. There was also the looming problem of declining reserves of clay that made the company even less willing to modernise the existing kilns.

Both Lidlington (which was normally called Marston works) and Ridgmont works were eventually closed by the LBC in the early 1980s due mainly to market conditions. Drivers travelling along the Ml in Bedfordshire no longer have the opportunity to attempt to count the number of chimneys that stood beside the motorway (there were in fact twenty chimneys!). During the mid-20th century the Marston Vale in Bedfordshire could boast of a forest of 108 kiln chimneys before rationalisation set in.

A later purchase by the LBC was the Whittlesey company whose origins go back to the early years of the century. They had been taken over by the NCB (who owned a large number of brickworks associated with coal mines) but in turn they offered them for sale by tender in 1973. The LBC bought Whittlesey in 1974 since the clay reserves in the Peterborough area could be better managed under the control of one company. There had also been a number of take-overs of smaller, non-Fletton, brick-makers, including Claughton Manor in Lancashire, Clock House in Dorking, Surrey, and Milton Hall in Southend. The theory behind these purchases was to reduce the company's reliance on the Fletton market.

All of these manufacturers produced high quality bricks in low numbers, but each brick was more profitable than its Fletton equivalent. With the Government pushing for new and cheaper materials for house building the Fletton industry was vulnerable to swings in the market. Therefore the company looked outside its own basic industry to attempt to diversify its interests; acquisitions of note included Croydex Ltd., the Banbury Buildings Company and Midland Structures of Bedford.

Top Right: *The first view in this trio of LBC constituent company vehicles is this Foden DG 8-wheeler ETM 472 of 1946 vintage. However, although it never became part of the LBC fleet, this wonderful vehicle with its attractive orange and black livery deserves a rightful place in this book. We now know, that more Marston's pictures have survived, and we would invite anyone involved with this firm to write to the publishers with their recollections!*

Centre Right: *The next Foden is HBM 74 in the fleet of H G Pentus Brown of Leighton Buzzard who employed a dark blue livery, whilst its regular driver was Fred Prentoe. It was contracted to Bletchley Flettons Ltd. but apparently conditions in their Bletchley Yard were quite deplorable.* Photo. Courtesy: P. Moth

Bottom Right: *Number 5 in the Eastwoods Fletton Bricks fleet, is a Foden DG6, which probably dates from pre-war days. This 6-wheeler was of the same type as the large batch of brick vehicles requisitioned by the Ministry of Supply in World War II. We now know that several of these Fodens were attached to No. 7 Co. REME and used in moving supplies to the Kent Coast to prepare defence installations after the fall of Dunkirk, others were used in aerodrome construction, while several AECs were handed over to the Ministry of Food Production and used to carry sugar beet.* Photo. Courtesy: P. Moth

THE LANDFILL OPERATION

One of the questions that to some extent dogged the company during the middle years of the 20th-century was what to do with the worked out clay pits. These pits were a serious eyesore and even when screened represented a waste of good farming land. The most recent pits, which had used mechanised clay-getting had almost vertical sides and could be up to 100 feet deep. Whilst some could possibly be filled with water (to at least make them look natural), they would have had no real use as leisure facilities as few people in the 1950s and 1960s had the time or money to use them.

One clever idea was to place new brickworks in the worked out pits themselves. The new works looked more like single storey factories than brickworks and featured just one or two very high chimneys to remove the waste gases. A major stumbling block to the agricultural renaissance of the clay pits, was the sheer volume of material required to bring the land back up to its original level. Meanwhile farming in the pit bottoms was not practical due to poor soil quality and frost pocketing. By the 1960s the sheer acreage of worked-out pits was becoming a problem. But at that same moment in time many of the Central Electricity Generating Board's coal-fired power stations were becoming embarrassed by the volume of wastes in the form of Pulverised Fuel Ash (or Fly Ash) that they produced. What better solution than to put the waste ash into the pits, then cover the ash with soil and allow the area to be farmed again.

Above: *One of the Landfill Nord-Verk off-road vehicles is seen moving an EASIDISPOSE container from the Freightliner terminal at Stewartby.*

This is exactly what happened in 1970 when the company set up a subsidiary called London Brick Land Development Ltd. whose main aim was to make the best use of the existing disused land. Their first success was to negotiate with the CEGB and British Railways to arrange to haul the waste ash in bulk, in the form of block train loads from the power stations to those empty pits, which had reasonably close rail links.

The dry ash had water added to it on discharge from the rail wagons to allow it to flow as slurry by means of a short pipeline into the pits. As the ash sank to the bottom of the pit the water was recycled to be used again to move more ash. The equipment was owned and operated by the CEGB with London Brick simply acting as landlords. For a long period this operation was very successful, but there came a decline in the volume of ash produced. The newer oil-fired power stations that came on stream produced less ash and an alternative disposal route for the ash was found in 'breeze block' manufacture, which incidentally pushed some Fletton bricks out of the internal wall market. In 1971 the Land Development Company purchased the Easidispose company who were based in Bletchley and operated a number of skip wagons for carrying waste and they became the nucleus of the fleet of waste vehicles.

Top Right: *An AEC Marshall PBH 294W fitted with a Mackrill body (and not Jack Allen as stated in the first edition of this book). It is identified by the rear-end lifter that could handle a variety of different size bins. Note the London Brick Landfill Ltd. lettering, which was carried on the cab side as the Easidispose name was not then widely known.*

Centre Right: *Within the waste transfer station at Hendon, three refuse collection vehicles discharge their loads. The two furthest from the camera are Seddon 16:4's both with the London Borough of Brent, but with different livery and lettering styles. The vehicle nearest the camera is the Borough of Barnett's HML 904K, a Shelvoke & Drewry T series refuse vehicle.*

Bottom Right: *These ex-RAF Leyland Hippo Tankers were employed on refuelling duties with the equipment that was used on the landfill sites.*

These operations were consolidated in 1977 when the parent company formed London Brick Landfill Ltd., at a time when a more profitable product was obtained to displace the fly ash in the disposal site. This new product was household and industrial dry waste that was available in ever increasing quantities from many areas including London, Bedfordshire and Northampton.

There was a shortage of tipping sites in the South East and a lot of waste was looking for a suitable hole in the ground to fill. The waste was present in copious volumes and the London Brick Landfill Company set up a trial operation with Northamptonshire County Council in 1976 to transfer waste from Northampton to Stewartby in road hauled containers.

At the old clay pit, now called the Landfill Site, the containers were then transferred by gantry crane onto articulated tipping trailers drawn by ex-army 6-wheel drive AEC Militants. During the next year the Landfill Company established a contractual partnership for a 20 year period with the Greater London Council. The agreement was to accept waste from the boroughs of Brent, Camden and Barnet via a rail connected transfer and compaction depot established at Hendon. This was beside the old Midland main line out of London, which went directly past the company's works in Bedfordshire.

The first pit to be used for both contracts was the 183-acre 'L' field at Stewartby, which was geologically suitable for waste. It had no fissures or cracks to allow any liquids to 'leach' away into local watercourses, and was close enough to the railway to enable the waste to be bulk hauled on special rail wagons. The contract required London Brick to build and operate a road/rail transfer station at Hendon in North London. At this terminal the incoming waste was tipped into special receiving bays and then compressed into special Road/Rail monocoque containers for the journey to Bedfordshire. The design for these containers (with a capacity of 20 tonnes or 28 cubic metres) is due mainly to a company director, Mr. R. A. Needs, as was the similar, but smaller terminal at Stewartby.

Top Left: *An EASIDISPOSE AEC Marshall TNK 466R, fleet number LD543. These vehicles were amongst the first to use the 'hook-lift' system for large skips. Better shown in the next picture, this system was a significant improvement on the wire rope system, and the driver did not have to leave the comfort of his cab.* Photo courtesy: P. Moth.

Centre Left: *This view of HMJ 239N on AEC Marshall LD535 shows how the hook-lift apparatus was employed to lift the waste disposal containers. Note the hydraulic rams which, although more costly, did not need the frequent replacement of the wire ropes.*

Bottom Left: *The EASIDISPOSE fleet operated by London Brick Landfill had a number of well-used vehicles. Pictured here this British Leyland skip wagon certainly looks to have earned its keep in the Bedford area. With the fleet number LD598, HMJ 354V is badged as a Clydesdale model.*

The containers were lifted off the rail wagons at Stewartby and transferred to specialised off-road vehicles to convey the waste-filled containers to the tipping face within the landfill site. The Northampton contract commenced in November 1976 and December 1978 marked the start of the London contract. In the following year the Landfill Company were able to accept some slightly more difficult wastes, including non-toxic liquids that were carefully introduced into trenches dug in existing wastes and then immediately covered.

A total of 160 acres of old clay pits at Calvert in Buckinghamshire were also used for household waste that was brought by rail from a transfer station in Hillingdon in West London. Later a similar contract was negotiated with Avon Council to bring waste by rail to Calvert. The pits at Calvert are situated in an area of woodland which forms the watershed between the Ouse and Severn rivers and the long term plan is to restore the woods and spinneys to the condition they were in before brick production commenced.

Further growth in the waste disposal field came in October 1983 with the purchase of the Dogsthorpe landfill site and a depot at Helpston, both near Peterborough. Both these acquisitions came from Biffa Waste, a company that had commenced in business as sand and gravel extractors and had themselves turned to the waste disposal business. A further purchase was Clearwaste from Thetford who were a subsidiary of Biffa and operated in the field of sewage disposal and its injection into (not onto) farmland as a nutrient.

The subject of waste disposal in its various forms, is a subject far too great to detail here, and this brief overview of the London Brick Landfill operation is just a brief overview of an area into which the company diversified in order to obtain additional revenue. The founding fathers of the brick industry and the early speculative house-builders in London would never have thought that the clay pits they dug to make the bricks, which built the houses, would one day be used to bury the waste produced in those houses that had been built in faraway London. For the full story of refuse collection vehicles, see the Nostalgia Road book *Municipal Refuse Collection Vehicles.*

Above: *One of the Nord-Verk off-road vehicles (fleet no. LD 4553) crossing a landfill site with an EASIDISPOSE monocoque container. This view is perhaps one of the least common to be found in a road transport book, as few photographers ever spent the day hanging round landfill sites. Yet, along with the picture below, it is a fascinating view of the difficult conditions that refuse vehicles had to contend with. Whilst these Nord-Verk units could take this type of terrain in their stride, the same conditions have to be faced by the conventional dustbin wagons (often based on commercial lorry chassis) that we see calling at private homes on a daily basis.*

Right: *Before the Nord-Verk tractors were purchased, the London Brick Land Development company obtained several ex-military 6-wheelers and converted them to haul the containers over the landfill sites. One example is AEC Militant LD536 seen here with a rather unusual Taskers trailer. When London Brick was acquired by Hanson, the Landfill Company was not considered a core business to manufacturing, Hanson sold off the operation, vehicles and landfill sites to Shanks & McEwan Ltd.*

THE HANSON TAKE-OVER

One of the biggest benefits of the semi-dry brick industry as typified by the LBC was the fact that huge quantities of bricks could be produced cheaply by use of the continuous kiln. This was very useful in the boom times, but could become a noose around the neck of the company when the government decided to 'turn off the tap' and reduce house building on the grounds of the prevailing economic situation.

There is no easy way to stop producing bricks and any fall in production has to be funded; add to this the rapidly changing market place and the need for much modernisation of plant and equipment and one will understand that the brick producing companies need deep pockets. The company had set up brick manufacturing operations in Australia and Iran to help offset the ups and downs of the British economy. The company was very lucky in making regular profits, but as they were finding some problems with funding these downturns, the operation was vulnerable to a take-over especially if the city considered that the return on capital could be improved.

This situation caught the eye of the financier James (later Lord) Hanson, who bid for the company in 1984 and possibly paid over the odds for it. At this time the Hanson Trust had a reputation for asset stripping and there was much gloom and despondency at the head office near Regents Park in London and in all of the works.

Above: Hanson's pedigree went back a long way, and they operated luxury coaches, haulage and parcels delivery vehicles and a number of decrepit service buses. This 1950s view shows a pair of the company's Maudslay Marathon's with Plaxton bodywork, fleet numbers 247-8 DVH 409 and DVH 682. The livery was bright red with gold lettering, but the white flash on the side disappeared in the late 1950s. (As an aside, we might record that one of Hanson's early take-overs was Earnshaw Haulage, following the death of the publisher's grandfather in the 1930s).
Photo: J. Haynes, courtesy Robert Berry Collection.

However, the outcome for the LBC was somewhat different to what many expected. What the Hanson Trust saw in the firm was not a 'lamb for the slaughter', but a company with a great potential for selling more bricks with greater profitability. He restructured the company by selling off some of the less profitable operations together with those companies not directly involved with brick-making. Although the volume of bricks sold never reached the giddy heights of 1974 when a total of 74 million bricks were produced in one week, brick production under Hanson became more profitable. Most importantly the improvements started by the research department in the 1950s and 1960s finally began to bear fruit with a superb and increasing range of modern, improved Fletton facing bricks and a modernised Fletton common brick.

Above: *The London Brick Company will always be synonymous with its fleet of well-kept AEC eight-wheel lorries, which they began acquiring in the 1930s. Despite losing most of these to Government requisition during World War II, a sizeable fleet was again built up in the 1940s and '50s. A survivor of the pre-war period is this Mammoth Major (BEW 605), which is seen here in its preserved state. All that is needed to complete the scene is a nice load of Fletton bricks.*

Right: *In addition to its fleet of 8-wheel AEC Mammoth Majors, LBC also had a sizeable fleet of 4-wheel Mammoths. As time went by the 4-wheelers were replaced by the AEC Mercury after that model was introduced in 1953. This was the first new lightweight truck to emerge from Southall after World War II, and as might be expected the LBC purchased large numbers. Here L182 (WMJ 973) is pictured making a delivery in London.*

Above: *With the advent of the Mercury, the LBC's smaller capacity fleet proved itself more nippy and able than ever before. It was not uncommon for these vehicles to do two or more runs to London every day. With the opening of the M1 motorway, it was found that the journey times were decreasing, and better ride qualities were introduced into the models being bought for the 1960s. Amongst the improvements was 'air suspension', which as the text describes was something of a mixed blessing. One of those with this kind of suspension was L150 (680 GTM), which was preserved at Mealsgate in Cumbria and is seen on display in Carlisle in 1996.*

Left: *The 8-wheel fleet also saw improvements, most strikingly in the design of the cab. These two Mammoth Majors were not originally part of the LBC fleet, as they were ordered by their rival (and later subsidiary) company, Marston Valley. They are pictured here at the MVB Ridgemont Works.* John E. Hardy

Above: *After its purchase of Volvo F86 and later F7 models, London Brick experimented with a number of 6-wheel lorries with 'Selfstak' equipment. For this they chose the Seddon-Atkinson 6-wheel 300 series model, which was fitted with an engine produced by International Harvester. At this time Seddon Atkinson were a part of the multi-national IH group, but the engine did not receive favourable comment by LBC engineers. By contrast the chassis and running gear were superb, so the company's 300 model lorries were something of a mixed bag, with some being better than others.* Seddon Atkinson Vehicles

Right: *The lorry park at the former Marston Valley Ridgemont Works, showing Volvos and 21 Ergomatic-cabbed AECs. One of these was CBM 242C, the sole vehicle of this kind bought by the LBC (the one with the headboard), the rest came in to the fleets via the Marston Valley company acquisition.* John E. Hardy

27

Above: *Many people will mourn the loss of the London Brick identity, along with the changes to the distinctive vehicle fleet that came when the new corporate livery was introduced by Hanson. Oddly Hanson, like London Brick had always operated a fleet of vehicles in a striking red livery, but as its association with the Dyestuffs Division of ICI grew, many Hanson lorries were painted in ICI blue. Perhaps it is no coincidence that the new corporate livery for Hanson Brick adopted a striking blue livery, although viewed as a whole it is also a very patriotic red, white and blue. Here a modern-day Volvo is pictured on a typical site delivery.*

Left: *The progress in loading and unloading modern brick delivery lorries is quite phenomenal, and no longer do loaders, drivers or site labourers have to 'hand-ball' the bricks. Here a Volvo, showing the London Brick side screens, is being loaded with banded and 'wrapped' pallets of bricks. Who said the old days were the 'Good Old Days'?*

PROGRESS UNDER HANSON

The company's biggest step was the introduction of the 'Kempston' brick in 1987, which was produced by a totally modernised Hoffman process. Although this was still a semi-dry brick, additives were introduced to give a moisture content of about 23%, these then combined to give a through-colour brick that was frost resistant. The clay is mixed in a series of single or double shaft mixers prior to entering the extruder. This machine works like a sausage-making machine extruding the mixed clay, which was then cut by wires to achieve the brick sizes. The green Kempstons are then fired just like the normal Flettons.

This new range of bricks met EC standards and proved to be a great success for the company. Meanwhile, the modern Fletton brick is much better in comparison to its counterpart from the 1950s. Since restructuring the company has gone from strength to strength, with a wide range of specialist bricks and fittings including for instance all the component parts to make brick arches on site.

For the sake of completeness, we might mention that Butterley Brick, was taken over by Hanson in 1968 and in turn the National Star Brick & Tile Co. was taken over in 1971. In 1972 seven brickworks were taken over from British Steel, six of which formed the Castle Brick Co. in Wales and the NCB's midland works followed shortly after.

With a large number of plants under their belt the firm was able to offer a wide range of bricks and successfully entered the Irish market in the 1980s. In 1982 Hanson purchased the Elland works of Samuel Wilkinson & Sons. The Butterley group initially concentrated on its traditional market of high quality facing and speciality bricks, whilst LBC stayed with the Fletton market. In 1999, the entire operation became known simply as Hanson Brick, and the individual names passed into history.

TRANSPORTING BRICKS

All of the brick companies in the Peterborough area relied on horse and wagon transport for brick delivery in the early days. With the bricks being very heavy and relatively cheap, the firm's main market was within perhaps a three mile radius of the works. The introduction of the traction engine would certainly have helped to widen the market for the bricks locally, but they were not really suitable for middle or long distance work, but steam power was available in the form of the railways! The Great Northern, had reached Peterborough during the period 1848-1852, and without a shadow of doubt the local stationmasters would have approached the brick-makers at a very early stage offering to supply rail wagons for the movement of bricks to distant destinations.

At this juncture it is important to note that at this time the road system throughout the British Isles was totally undeveloped. There were few lengths of properly metalled roads outside of towns, and generally only cart tracks existed elsewhere. These were either impassable in winter or dust bowls in the summer, and usually difficult to traverse at any other time. In this situation the railway companies had very little competition to their goods delivery service to both the capital city and to many other centres of population.

The potential of the safe, quick and reliable carriage of bricks to their customers must have seemed like a miracle to the owners of the works. Even so, in the first few years of brick delivery by the independent railway companies it would have been necessary for the bricks to be taken to the nearest railway goods yard by horse and wagon.

This operation called for the bricks to be manually loaded onto the rail wagons and the same process would be repeated at the delivery point with the customer being responsible for removing the bricks off the railway wagon. Once the Great Northern Railway realised the potential volume of brick delivery business that could be developed, they very soon started to build private delivery sidings to serve the brickworks and arrange storage facilities at their goods depots at the major destination points. The sidings enabled the brick-makers to load directly into the rail wagons from the kilns or from the stockyard. It was probably a two-way decision since the more astute brickworks owners made sure that their new kilns and works were located alongside a railway line. At a very early stage the vast majority of works, even very small ones, all had their own sidings. Only the Norman Cross works (close to the A1 road) was known to be lacking in rail access, so horses or traction engines were used to transport the bricks to the railway.

Left: *A typical view of a London Brick long-distance lorry is seen with F106, an 8-wheel AEC Mammoth Major HBM 831, which is loaded with clay blocks. These blocks were a development from agricultural drainage tiles, but in the post-war era they were widely used for building internal walls prior to the advent of plasterboard.*

Top Right: *For its local delivery work, the Marston Valley company operated a fleet of Bedford O-Types, as seen in this view dating from about 1954.* Vauxhall Motors

Centre Right: *Rough shunting of rail wagons (such as these owned by the LMS and GWR) could cause immense damage to a large quantity of bricks that were often urgently required by the customer. In the case of urgent requirements for the delivery of bricks to an out of the way address, there was no quick way to expedite delivery.*

Bottom Right: *Many innovative ideas were tried out to reduce the incidence of damage including the introduction of air-filled dunnage bags that were made of rubber. Since the 'Pipe-Fit' open rail wagons used for brick carrying worked only for London Brick on specific closed circuit working, the bags did get returned! Perhaps the final straw though was the railway's insistence on increasing transport charges, which had the effect of making bricks more expensive. This 1970s view shows the Freightliner bay at the Stewartby works prior to the commencement of landfill operations, which utilised the same sidings (see page 20).*

At a very early stage the vast majority of brickworks, even very small ones, all had their own sidings. Only the Norman Cross works (close to the Al road) was known to be lacking in rail access, so horses or traction engines were used to transport the bricks to the railway. One yard in Whittlesey also boasted of being the only one connected directly to the canal network! But it was the railway network that became absolutely vital to the fledgling Fletton brick industry.

With the opening of the Great Eastern railway to Whittlesey the companies had access to the eastern suburbs of London, in addition to those services already offered by the Great Northern Railway to the northern suburbs. The Midland & Great Northern Joint Railway (often known as the Muddle and Get Nowhere Railway) could offer services into rural East Anglia, serving the agricultural areas where a demand for brick was steadily growing. The goods agent at Peterborough Central handled all administration work with assistance from staff at Yaxley and Whittlesey goods yards.

In the Bedfordshire area the Midland Railway from St. Pancras to Derby, bisected the clay bearing lands. There was also the Oxford to Cambridge route, which had opened between 1846 and 1851, and (much later) the Great Central Railway from Rugby through Woodford Halse and the Vale of Aylesbury. At various points along these lines a number of brickworks were established, and the most significant of these was that at Calvert where the Great Central crossed the Bletchley to Oxford line.

Top Left: *The company's first delivery vehicle, was a Morris Leader with the registration number EG 2207. A batch of similar vehicles was purchased in 1938, before allegiance was transferred to AEC. One of the 1938 acquisitions was AEW 515 pictured here.*

Centre Left: *In July 1935 a decision was taken to employ a full time professional manager to run a transport department which was to be set up at the Fletton works in Peterborough. It commenced operations in February 1936 with 16 cars, eight trucks and a 20-seat bus.* Courtesy: Phil Moth

Bottom Left: *Later that year they purchased their first heavyweight vehicles, 11 Leyland Cubs, and a mixed fleet of 25 AECs including 4-wheel Monarchs and Mammoth Major 4- and 8-wheelers. Including the 4-wheel Monarch EW 9245 seen here. This was really the start of the long-term relationship with AEC, but it was an arrangement that all were to benefit from.*

Since the majority of bricks in those days were destined to supply the needs of the speculative house builders in London and the south east, it was vital that the production costs of the bricks and their transport charges were both kept to a minimum. This was especially important if the makers wanted their Fletton bricks to compete with bricks produced locally in the London area. Minimising transport costs was seen to be an important factor to Mr A. W Itter's search for new clay pits when he bought land adjacent to both the new GCR and the LNWR lines. This allowed him the opportunity to play one railway off against another, for the railways quite often dramatically reduced transport rates when they faced competition for traffic.

During the early years of the 20th-century, the railway companies moved over 90% of Fletton bricks with the balance being delivered locally near to the works or handled by third party contractors (i.e. road hauliers). By this time most brickyards had built loading platforms at the height of the railway wagon floors to obviate the need for ramps and barrows to move the bricks from ground level. However, despite the sheer volume of bricks being moved by rail the overall position was not really satisfactory.

In the first instance the brick companies had no real control over the final delivery of their product. The bricks may well be unloaded by unskilled labour into unsuitable vehicles at the delivery depot, with multiple handling causing damage and shortages. Secondly, even though the railway service was comprehensive in that most towns and villages had their own goods depots no one could be quite sure when the rail wagons would arrive. Whilst this was not desperately important at the larger terminals where stocks of bricks were maintained, it could cause problems at smaller depots. Thirdly, the brick-makers were concerned that, following the mergers between the larger companies to form the London Brick Company, their production facilities might soon outstrip the railways' facilities to handle them. The increased production also gave the company the ability to sell bricks into towns where it had never previously been considered to be economic.

Consideration of these adverse aspects of rail-delivery suggested that a company-owned fleet of road delivery vehicles would be a great asset, both to show the railway companies that they did not have a monopoly and also to offer a better overall service to the customers. Right through the growth years of the Fletton brickworks and up to the 1930s, the railway companies (by then merged into the London & North Eastern in the east and the London, Midland & Scottish in the west) still carried the majority of Fletton bricks. They certainly had no monopoly with many of the manufacturers using third party road hauliers to keep the railway companies on their toes.

Few manufacturers operated road delivery vehicles of their own possibly on the basis that transport was best left to the professionals. Despite the hesitation to get involved in the delivery of bricks by road the London Brick Company did dip their toe in the water by purchasing a variety of vehicles from 1928 onwards.

With true British aplomb the very first vehicle was a Morris Commercial 1.5 ton truck. After much thought another truck entered the fleet in 1930, this time an Albion, which was then followed by a Manchester bus (which may well have been purchased second-hand). A brief note on the Manchester make might be appropriate here as the Crossley company began assembling American Willys-Overland vehicles for the British market and later sold them as Manchester's until this operation went into liquidation in 1933. A small number of Morris and Bedford trucks then joined the fleet for use in the internal works fleet and possibly for delivery. One well recorded use of the firm's road vehicles was the delivery of bricks to the Royal Estate at Sandringham.

Top Right: *The 8-wheelers purchased in 1938 weighed an incredible 6 tons un-laden and as they were fitted with the 7.7 litre diesel engine they gave a payload of 16 tons. These 8-wheelers came fitted with 'two stick' transmission, that is two gear levers giving 5 gears, 1-4 in low ratio and 4 high. The LBC continued with AEC 8-wheelers until the 1960s, as can be seen from this picture of the fleet at rest.*

Centre Right: *The AECs all had diesel engines and were members of a standardised range of models introduced in the early-1930s by their chief engineer Mr. J. Rackham. These new models were loosely based on the earlier heavy and over-engineered Mammoths, but had been totally re-engineered to reduce weight and production costs to enable them to compete with their Lancashire-based rivals.*

Bottom Right: *The LBC's Leyland Cub trucks all had Duramin alloy cabs and bodies like the AECs, as the use of such bodies was intended to contain the un-laden weight to below the magic 3-ton figure and thus enable the trucks to travel legally at 30mph. Originally fitted with petrol-engines, a number of the Cubs received diesel power units two years later at a cost of £125.6.9d (£125.34) each, which certainly would have increased the un-laden weight. These Leylands were, like the Equiloads, used generally for deliveries within a 30-mile radius. Interestingly, the remaining petrol-engined Cubs left the fleet before the war, but most of the original AECs lasted until the late 1950s.*

Above: *One or two of the 8-wheelers appear to have been used as mobile gun emplacements judging by photographic evidence. Even the original AEC Mammoth Major 8 (F1, EW 9246) seen here, which had been regularly driven by Bill Frisby was taken away, and never returned, so the fleet number was used again. Among the impressed vehicles was almost the entire batch of 1939 Morris Equiloads none of which ever returned.*

Other purchases in the 1930s included a small number of commercial vehicles including Bedford Unipower tippers and a Morris-based fire tender for use within the works. With the ending of the Depression in the early 1930s, the company considered setting up their own transport fleet to supplement and improve upon the delivery service offered by the railways and hired vehicles.

The fleet strength grew in proportion to the brick sales during the late 1930s and by 1939 there were 56 Leyland Cubs or Lynxs, 238 AECs, 38 Morris Equiloads and Leaders, four Bedfords and two Thornycrofts in operation. Of this fleet the Morris trucks had been chosen because they were British and considered relatively heavy duty. The Leyland Cubs offered a high quality, hand-built vehicle able to travel at 30mph due to their low un-laden weight. The AECs although heavier and restricted to 20 mph were obviously totally satisfactory in every way.

By 1939 the company were manufacturing bricks at a total of 27 works, but the vehicle fleet was maintained at just five depots; namely Peterborough, Bletchley, Arlesey, Stewartby and Calvert. The main offices and works had been established at the Fletton works, where a brand new heated vehicle workshop was built to replace an earlier open-air maintenance facility.

In those days engines were de-carbonised at 10,000-mile intervals and major overhauls took place every 50,000 miles. One fitter recalls that the major overhauls were just that, as the vehicles were stripped right down and rebuilt with as many new parts as deemed necessary; a large stock of AEC parts was always kept in store to meet any eventuality. A great deal of 'other work' was carried out at the Fletton workshops including the conversion of two 1936 AEC Matador 4-wheelers to Mammoth Major specifications. Both vehicles had been bought second-hand and as ex-works demonstrators. These 'new' vehicles then received 1938 registration numbers BEW 119 and 120 and remained in service until 1960! The conversion cost was £887 each making the overall price £200 more than a brand new 8-wheeler.

There was a decline in brick sales in early 1939 so the company, obviously thinking ahead, promoted the building of brick air raid shelters to the public. Prior to selling the idea of brick shelters, the company's research department under Mr. T. Boxall even tried to blow up a prototype shelter built at Stewartby to prove how safe they were. By the summer of 1940 Government restrictions on anything but the movement of essential goods had curtailed further sales of bricks.

Thereafter it was the placement of huge orders by the Ministry of Works for the supply of bricks for airfields, office blocks and war factories that really counted. The decline in general sales, coupled with the loss of skilled men to the armed forces, significantly reduced production capacity. The company took steps to alleviate the labour shortage by employing women in many posts within the works, but despite the undoubted capability of the girls employed in such hard work, the sales of bricks continued to decline. Consequently, several of the smaller works were either closed or mothballed.

Despite the brickworks operating below capacity, the workshops never had enough hours in the day. The engineering works with their highly developed skills in building and maintaining brick-making machinery were busy on manufacturing components and parts for the war effort; they even assembled and road tested tanks and other armoured fighting vehicles that had been shipped across from America.

Not only were the staff involved in the war effort, but to the initial surprise of the company's management a number of their commercial vehicles came to be requisitioned by the military authorities. A total of 57 vehicles were 'borrowed' for His Majesty's Service. Ostensibly they went for the duration, but only a few were actually returned to the company during the latter part of the 1940s. To replace the impressed trucks the company had to resort to purchasing second-hand trucks, a number of which entered service in 1941.

Amongst this little fleet were some Leyland Octopuses, an articulated Leyland Beaver and an ERF 8-wheeler. Yet, someone in Government obviously took offence at the LBC's ability to replace missing vehicles and promptly impressed eight of the 19 second-hand trucks, none of which ever came back. Little is known of the exploits of these vehicles after they left the LBC, except for one incident when a Leyland Cub caught fire whilst carrying a load of bombs!

Top Right: *This Leyland Lynx was registered AEW 787 and on entry to the fleet in 1938 it was given the number B51. One of a batch of 20 similar trucks it had a long life and lasted until 1956.*

Centre Right: *An undated photograph showing builder's and merchants' lorries collecting bricks from the works. Leonard Mascapelli writes, 'Saturday morning work was always available for loading loose bricks, and merchants sent lorries to purchase loose bricks at a fraction of the normal price.*

Bottom Right: *This AEC, registered BEW 73 was originally purchased as a 4-wheel Monarch, but it was later converted to a 6-wheel Mammoth with a 3-way tipper body.*

POST-WAR GROWTH

Fifteen of the impressed lorries were returned after the war, including the first AEC Monarch. Although the authorities did pay LBC for the vehicles that they 'borrowed', there was a sting in the tail. The Morris Equiloads had cost the LBC £336.10.0 each and (at three months old) were bought for £303.0.0. The AECs were around three years old when called up in 1942 and even though a new Mammoth Major 8 cost £1,685.13.6, the authorities only paid £781.0.0 for BEW 100. Yet, when it was returned after the war the LBC had to pay £1,048.1.3 for the privilege of buying it back!

All vehicles were of course in short supply at that time and the company was desperate for delivery vehicles to meet the heavy demands for bricks. Although the company records are very precise for the wartime period, only one AEC G65 was noted as having been converted to run on producer gas, but no doubt other vehicles were also involved.

The wartime restrictions on transport had severely reduced the delivery area for the bricks. In addition the production of bricks was in turn restricted by a quota system, which in turn resulted in a stockpile of 500 million bricks; yet these had all been sold by 1947!

Above: *A Morris Commercial Equiload, EBM 27 with fleet number DP84 was purchased on 4th March 1946, and withdrawn on 15th February 1955. The Equiloads, although quite competent, were soon relegated to works use including this example being allocated to the 'Stores Department'.*

This amazing feat was achieved in spite of the atrocious winter in which coal shortages had restricted production and freezing weather caused total chaos on the roads. The company then built up production with the help of a night shift, only to find that the sterling devaluation and other cuts forced on the Government in late-1947 reintroduced the spectre of stop - go policies to defend the British economy. This deflated demand and forced house building to stop virtually overnight.

Of course this was only to be one of the many temporary blips in the history of brick production and by mid-1948 the demand had returned, but the company was by now short of labour. During the conflict a number of prisoners-of-war had served the company, but most had been repatriated and refugees from Eastern Europe had taken their place. But labour was not the only thing in short supply, as new vehicles were almost unobtainable.

Top Right: *As a result of post-war vehicle shortages, a fleet of 17 ex-War Department Chevrolet 4 x 4 CMP chassis were purchased and fitted with tipper bodies for internal use within the brickworks. Whilst they were very cheap to buy, they were certainly not inexpensive to maintain and because of this they were replaced by Ford diesel tippers in 1953.*

Centre Right: *The LBC purchased a brace of ex-WD AEC Matador 4 x 4's mainly for use as breakdown vehicles. However they also became heavy haulage units when combined with low load trailers to move large and heavy items of plant.*

Bottom Right: *To expand capacity, the LBC workshops converted the remaining AEC Mammoth Major 6 platform trucks to 8-wheelers. Amongst the vehicles altered in this way was G8, the preserved BEW 77. The G series of numbers then became vacant and they were used for the new Albions then being acquired. Further purchases of AECs in the period included the 4-wheelers like D225 (EBM 121), which is seen here fitted with drop-sides for specialised work.*

Despite the shortages in the first full post war year of 1946 the LBC was allowed to order some new lorries, and it actually obtained them remarkably quickly. A further 34 Morris Equiloads thus joined the fleet along with 35 AEC Monarchs and Matadors. The latter were the post-war haulage model type 0346 (rather than the wartime 0853, 4x4 models) and came fitted for the towing of a batch of 28 Taskers drawbar trailers. The Monarchs were used as solo units and had the model designation 0346S. These trucks had a single front engine mounting with the rear being steadied by a torque bar. But when the bushes became worn, the engines would rock from side to side leading to several drivers being stopped by the police and told that their engines looked as if they were about to fall out.

In fact these Monarchs were colloquially known as 'Swingers' or '4 Oilers'. Despite this and other minor faults such as a penchant for breaking crankshafts they lasted until 1965 when they were sold at auction for between £57 and £95 each! AEC continued to supply a large number of vehicles including Mammoth Major 8's until 1950, but around this time AEC began to fall behind with its supplies, so the 4-wheel fleet was supplemented by a large number of Albion Chieftain FT37 platform trucks and tippers. These tippers, the first big purchase of this type in the fleet, served a dual purpose in delivering bricks to the East Midlands, then returning with coal for the kilns. The bricks were loaded by hand in the normal way, but on arrival at the building site the bricks would simply be tipped out onto the ground.

Whilst this might seem an alarming practice, a skilled driver could in fact deliver a load of bricks and cause less damage than would be the case if they were 'hand-balled' off the wagon. As a matter of fact, in many areas, for a long time it was common practice for bricks to be delivered by this method. It must be pointed out that only common bricks arrived on site by tipper; facing bricks were always given the courtesy of personal attention from the site labourers.

Vehicle maintenance by this time had returned to peacetime standards, with routine servicing carried out at night and scheduled overhauls and other repair work being handled by the day staff. With the introduction of better oils and improved fuels, decarbonising had been extended to 25,000-mile intervals.

The AEC 7.7 litre engines fitted to large numbers of the fleet would run up to about 80,000 miles, after which the engines would be exchanged for reconditioned ones. During the war there had only been minimal regular maintenance and with the lack of air cleaners, no hard valve seats and no shell bearings, this mileage was about the best one would expect.

As better quality vehicles (and engines) entered the fleet, overhauls were stretched to 50,000 mile intervals when pistons were generally replaced. Then, at 100,000 miles, engines were totally rebuilt. At these major overhauls the vehicles were completely stripped down to their component parts. Even the chassis frames were dismantled, although generally only the odd cross-member needed replacing. The actual rebuilding work required the services of the plant maintenance fitter to re-rivet the chassis!

Above: *One of the big purchases in the post-war era were the AEC Mammoth Major 8-wheelers, of which F95 GTM 809 is just one example. It was acquired in August 1949 and lasted until the autumn of 1963. In all 14+ years of repetitive hard work, hauling the none too light loads up and down the country's still inadequate road system. The longevity of these trucks is therefore a tribute to both AEC and the maintenance engineers of the London Brick Company.*

The company undertook all work in-house including rebuilding gearboxes, rear axles, brake components and steering gear so it held these items in stock. In those days all components were rebuilt on a strict mileage basis regardless of the fact that some items like AEC differentials might run forever. The bodyshop, in addition to repairing accident damage, also rebuilt cabs in their spare time.

Amongst the jobs undertaken by the workshops was the conversion of various petrol-engined vehicles to diesel power, including the fitting of Perkins diesel engines into the Morris Leaders.

Despite the ostensibly better quality of the pre-war Leyland Lynx's they regularly came to visit the workshops, not through breakdowns, but in order to get the rear brakes better balanced. The brakes had a combination of hydraulic and mechanical actuation and were the cause of much heated comment within the workshops. The Chieftains, which eventually replaced the Lynx lorries, also had their own little foibles. For instance, they gained a rather bad reputation for the considerable difficulties encountered in setting up the induction butterfly valve that helped create the vacuum for the braking system.

Haulage work was not limited to just the LBC fleet, as private firms also found it very easy to 'get a load' at the works. One of the regular hauliers was Cadelent's of Toddington in Bedfordshire. According to the well respected transport historian Peter Davies, this company was well known for their fleet of ex-army vehicles; some of which were Bedford QL (4 x 4) models that had been converted to single (rear) drive with Thornycroft front axles in place of the driven Bedford axle. The Bedford petrol engines were replaced by Perkins P6 diesel engines and the chassis were shortened, fitted with fifth wheel couplings and hitched up to ex-US Army single axle trailers. They made quite a spectacle and were not alone amongst the 'bitza' trucks operated by the hard-pressed hauliers calling at the LBC during the early 1950s.

Top Right: *Despite the seeming normality of vehicle deliveries during the early post war years quite a wide variety of less common vehicles entered service having been purchased for specific purposes. For example a batch of ex-WD 4x4 AEC Matadors were obtained. With the introduction of these large vehicles the fitters became responsible for all breakdown and recovery work. One of the regular calls was to put members of the tipper fleet back on their wheels after they had fallen over whilst tipping!*

Centre Right: *A very trying time was the winter of 1946-7, when the weather caused untold havoc. The picture of an unidentified 8-wheeler stuck in snow, will give some idea of the work that would have to be undertaken to get things moving after the wagons were thawed out. It usually required a new radiator to say the least!*

Bottom Right: *A number of contractors regularly supplied transport to the LBC especially for the distant destinations where it was not profitable to send company vehicles, as these operated on 'C' licenses and therefore could not carry return loads. Furthermore great use was also made of 'spot hire' hauliers and return loads to most parts of the country were given to any haulier calling at the brickworks. Here we see a BRS Austin K4 (GRH 236) being loaded with bricks destined for Yorkshire. This was not a one-off load as the first edition of this book stated, but was part of a regular contract taking bricks to the Drax Power Station construction site. The regular driver of this lorry was Godwin Hughes, and he writes:- 'The collection work was based on one return trip a day, but by returning to load up after the trip to Goole, seven trips a week could be achieved. This meant £3.2.6d extra a week, which over eight weeks made it well worth the effort.'*

39

Top Left: *Vehicle purchases settled down during the early 1950s with orders placed at Southall for AEC Mammoth Major 8's and at Scotstoun for Albion Chieftains, until 1953/5 when a small batch of Leyland's joined the fleet. The Leyland's comprised of Octopus 8-wheelers, model 22.01 and 24.01, (the first two digits representing the gross vehicle weight). The second batch echoed the increase allowed under the Construction and Use Regulations. These Leyland's were fitted with double drive and would go virtually anywhere, although the fitter's despaired at changing the rear springs after the drivers had made use of the go anywhere facility. Despite their many advantages these lorries only lasted until 1967/68. This was relatively short by LBC standards, but it may well have been the introduction of the Plating & Testing regulations that hastened their demise. This small fleet's main contribution to the LBC, was the ability to extend major overhauls to 150,000 miles, which of course gave major cost savings.*

Centre Left: *Although the Leyland Octopus 8-wheelers had been the first batch of vehicles from that make to join the LBC fleet since pre-war days, the company had acquired a solitary Leyland Comet registered in Lancashire with the number RTC 820. It was one of the few commercial vehicle purchases in 1953, but this 4-wheeler arrived as a demonstrator. After evaluation by the company in 1953 it was later considered worthy of purchase. Indeed, it was noted at the time as being the fastest vehicle in the fleet and it may well have been the precursor of many more had it not have been for the arrival of a new AEC model. One driver was Peter Harris, who recalled that this vehicle could do almost anything that was asked of it except when it came to hills with a full load. At the time he was driving the Comet they were doing a lot of runs into Coventry, but on one occasion, the Leyland's brakes failed near Crick, and the loaded lorry crashed into the parapet of a bridge on the Grand Union Canal.*

Bottom Left: *It was the introduction of the AEC Mercury in 1953 that persuaded the company to purchase AEC 4-wheelers again for use as maximum capacity two axle vehicles. The new engine in the Mercury was more powerful than that in the Comet and the truck weighed over a ton less than the equivalent Monarch. It also benefited greatly from the new styled cab, which featured an enclosed radiator and also an air cleaner that reduced engine wear, whilst the engine itself was a great improvement on previous AEC units. At this time AEC still relied upon outside coach-builders to supply cabs, though of course many chassis had Park Royal built cabs. The LBC turned to Duramin to supply light alloy cabs and platform bodies, giving a further weight saving. Whilst the Mercury had the benefit of new engineering, the other models in the range continued to look much as before although the Mark III range could boast new engines and axles as against the Mark II models, and came fitted with full air brakes. Towards the end of production, the Mark III Mammoth Majors actually sported enclosed radiators on the older design chassis. Pictured on delivery, and still sporting trade plates, is the air-suspended No. L1 (RNM 401), which arrived in July 1956.*

40

The company was never shy of trying new ideas that might help productivity or reduce damage to bricks in transit. One idea that seemed to herald a new approach to the question of ride quality was the introduction of rear air-suspension on 4-wheelers during the late 1950s and early 1960s. Taken overall these early air suspensions were generally reliable and did possibly reduce load damage. However the suspensions suffered from levelling valve problems often leaving one side higher than the other. In fact setting the valves could often take longer than changing a rear spring! There was a regular problem caused to the air suspension where stones would get caught in the convoluted rubber bellows and wear holes that then leaked air.

Any chassis fitted with air suspension was a weird sight whilst being unloaded by hand from one side, as the vehicle took on an almost nautical lean. The rubber bags would deteriorate if left without air over a period of time and could leak when re-inflated. When left without air the vehicle would 'settle' down until the wheel arches rested on the wheels and the air system was set up so that the bags would fill with air before the brake reservoirs filled! Looking back, the air suspension proved really no better or worse than the normal steel springs, especially if one considers that drivers of these vehicles might deliberately try to drive carefully when fully laden to cause less damage. When empty these vehicles certainly gave a better ride, but eventually economics ruled and later orders called for semi-elliptic springs on the rear suspension.

Top Right: *For quite a long time Duramin had supplied light alloy cabs and bodies to the LBC, as for example seen on this short wheelbase Mercury (VMJ 25) in 1958. Various corrosion problems (caused by a too high copper content in the alloy floors) and an unhelpful attitude by Duramin during the late 1950s, made the LBC consider the growing trend to use fibreglass for future vehicle cabs.*

Centre Right: *Many of the LBC's new fibreglass cabs featured forward-opening doors, which gave better access for the driver when compared with the standards set by the normal AEC/Park Royal or Duramin cab. The new cabs were built by Road Transport Services, and one is seen here fitted to Mammoth Major 511 BMJ. These cabs did not rot, but they remained colourfast and were easier to repair than the alloy equivalent. A regular problem was experienced with the door hinge mountings coming loose, but this was eventually rectified. The RTS bodies were designed with thicker head and tailboards, along with a much-altered body mounting on the chassis. These (and other) improvements made the bodies less prone to damage by enabling the body and chassis to move relative to each other.*

Bottom Right: *Originally fleet no V9, 715 GTM is a 1962 AEC Mercury that was fitted with a tipper body for the joint function of brick delivery and coal collection. Peter Harris also had a time driving this vehicle, and recalls that one of its frequent trips on the runs to Coventry involved a return load of coal from Daw Mill Colliery.* Photo Courtesy: Phil Moth

Top Left: *As the 1950s drew to an end, the London Brick Company was still placing a good deal of business with the AEC company in Southall. One of the last batch of vehicles purchased that decade included WMJ 988. Pictured here in March 1963 bearing fleet number L117, the Mercury has been fitted with mesh sides and tail boards. When bricks were hand loaded onto the lorry bodies there was little chance of them coming loose, so the benefits of good load security were well known to the company. However, once vehicles were released from the 20 and 30 mph restrictions imposed by the Ministry of Transport the situation changed somewhat. Whilst an individual brick may look quite innocent when sat securely on a lorry platform, once it slides off the vehicle on a corner it can become a dangerous projectile. Therefore a move was made to jig-build hinged drop-sides from Duraluminium. These mesh drop-sides certainly worked, but unfortunately proved very susceptible to damage and they frequently jammed shut. Their demise came because they were not fully compatible with Selfstak unloading systems.*

Centre Left: *In the early 1960s the AEC company at Southall were producing at full capacity and apparently unable to meet the requirements of the LBC. As a result of this there came a major break in tradition, as the company placed orders for a fleet of LAD cabbed Albion CH3 model Chieftains. Mostly rigid wagons were acquired, but two tractor units were purchased for use with Taskers 12-ton trailers, including 501 EBM seen here. The Albion vehicles were renowned for their cool running and invariably operated with the cooling fans disconnected. Yet despite this simple and cheap modification there was still excessive piston-wear. Mechanically they were well thought of overall, but the rusting cabs on many of these vehicles prompted their early retirement.*

Bottom Left: *The final AEC Mercury was a very early Ergomatic cabbed model and remained in the fleet for a long time whilst various experiments were carried out on it. The general opinion of the Ergomatic cab was that it was well thought out, quieter, more comfortable and certainly less cramped than the earlier AEC/Park Royal offerings. Certain drivers considered it was not really a major advance over the RTS fibreglass cab, which is strange when one considers how much time and money the Leyland group spent in getting this cab designed to suit the driver. In passing it is interesting to note that even in the 1990s the Ergomatic cab is still considered to be well designed and easy to drive, let down only by the noise emanating from the engine. Mention of the driver's view brings to mind the comment that the drivers apparently divided into two categories; there were 4-wheeler drivers and 8-wheeler drivers and when they stopped for breaks at the local transport cafes, the two groups kept to separate tables! Compare the picture of CBM 242C seen here with the modified vehicle shown with 'loading equipment' on page 39. It is pleasing to say that pictures of many of the vehicles shown in the first edition of this book were used by Corgi, as part of its range of 1:50 scale models.*

Top Right: *Corgi have also recently introduced a number of 4mm scale models in its Lledo 'Trackside' range. With respect to the prototypes on which the models are based, the company expended much time and effort in perfecting a load protection system that was effective in use, simple to apply, not easily damaged and able to withstand site abuse. From fitting stronger headboards the next move was to use rope netting bought from fishing-fleet suppliers Cosalt in Grimsby. By 1966 the company had come to the definite conclusion that the smallest delivery vehicles were to be 6-wheelers with a carrying capacity of about 7,000 bricks, and as a result a large fleet of AEC Marshal's entered service from this time on. These Marshals were in the main lightened versions of the standard model in line with LBC policy of maximum payload. Unfortunately delivery delays, the standard of build, the quality of service back-up, and spares supply, were all beginning to falter following the closer integration of the Southall company with its former Leyland competitor.*

Centre Right: *After the problems with AEC it therefore came as no surprise when the first Volvo F86 6-wheeler entered the fleet in 1971. Fitted with a luxurious continental style cab, lifting rear axle and a quiet, powerful turbo-charged engine this new type of vehicle delighted the drivers by making their job a little easier. Little things like the insulated and soundproofed cab, a suspension seat and air-assisted clutch were much appreciated by the drivers. There were a few teething problems with these vehicles, the most noticeable being the propensity of the turbochargers to lose their blades that quickly dropped into the engines causing serious damage. The Volvo fleet was not totally trouble free, for instance they had oil leaks and transmission troubles, but these problems were of small significance compared with what went on before. The foundation brakes were never as good as those fitted to the AECs with cracked drums becoming a recurring problem, but the overall package was excellent and running costs were again reduced.*

Bottom Right: *The most significant point to note about the Volvos was that (come MOT test time) the fitters check sheets would be blank compared with lots of X's on the Leyland group vehicles. The F86s were partly responsible for extending the major overhaul periods to 200,000 miles from the 150,000 mile schedule set by the Leyland Octopus 20 years earlier. Not to be outdone, some of the AEC AV505 engines in the AECs could run up to 400,000 miles as long as the pistons were changed at 200,000 miles. One impressive statistic was that Phil Baker the transport engineer was regularly able to offer a 97% fleet availability in 1974. This imposing picture shows the fleet 'at weekend rest' in the mid-1970s, when they were all kept in their 'types', and at least 15 AECs, 14 Fodens and 11 Albion's can be seen. It is said that the Albions would 'roll' considerably when loaded with hollow bricks from the Clock House Works and, as a consequence, were often known to 'shed' their loads. The picture also shows 42 Volvos and we wonder if Corgi will bring out a model of the LBC F88, now they have produced tooling for this vehicle type.*

MATERIALS HANDLING IMPROVEMENTS

Like many other companies the LBC treated the transport department as just a service to the sales and marketing department rather than as a stand-alone cost centre. As a result costs did not get monitored as closely as they should have done, with customers needs outweighing overall efficiency. From 1968 the department was asked to both reduce costs and improve productivity in order to increase service levels to all customers, not just those that shouted loudest.

There were a number of alternative ways to increase productivity: larger trucks could be employed, but articulated trucks would have access problems on many sites. Double shifting vehicles was a simple answer, but building sites only worked single shifts. Alternatively the drivers could be encouraged to drive a little faster in view of the increasing mileage of motorways coming into use. Or the time spent loading and unloading vehicles could be reduced!

Strangely enough the sales department always wanted the smallest possible vehicle to enable easy access to sites, but when one considers that the average house in that era took around 25,000 bricks one can understand the transport department's need for the largest possible vehicle. The company had very successfully introduced mechanical handling of the bricks within the works already, so could this progress be extended to the vehicles?

Above: *The traditional means of delivery as viewed with this AEC Mammoth Major HBM 837 of 1950s vintage.*

Below: *An advance in unloading came with road head distribution, but as this picture shows, its loading ramp was still far from satisfactory in terms of operator safety. However, note the cargo safety net on the Albion's load.*

It was obviously possible to make units or blocks of bricks for loading onto the delivery vehicles by standard fork trucks, simply by steel banding the bricks together and then using a standard hydraulic lorry crane to unload the bricks. In fact the Building Research Station in Hertfordshire had already designed a banded 50 brick 'pack' for use in the restricted areas of London. In Surrey the Redland Company had introduced the Dutch HULO trailer that could unload bricks by mechanical means, but this was very expensive and really only came in articulated self-steering trailer form.

The management wanted a total mechanical handling of bricks, but the hydraulic cranes of this period were too heavy for mounting on a road vehicle and not totally reliable. Furthermore, the cranes that were available at that time were to some extent over specified for a job that simply required the bricks to be placed beside the vehicle. After much thought and design, plans were drawn up for a very basic gantry type crane fitted with moveable arms. This original crane was nicknamed 'The Rocket' and had been the brainchild of the LBC's Transport Development Manager, Bill Needs, who was an ex-London Transport apprentice and had also worked for AEC.

Top Right: *The original design of brick unloader was fitted to the company's one and only Ergomatic cabbed AEC Mercury CBM 242C of 1965 vintage. On this flat-bed, the head and tailboard were altered and the hydraulically operated arms fitted inside them, the arms would unfold and project over the body side. A longitudinal beam carried a trolley that would lift the bricks in packs of bricks banded together.*

Centre Right: *The first crane experiment soon gave way to a modified version built by Motorail of Bedford and fitted to this LAD-cabbed Albion, it proved quite successful although a little over-sophisticated. The overall success of this first attempt at mechanisation was somewhat tempered by persistent failures of the steel banding holding the bricks together. It was then discovered that a Canadian company were manufacturing a system called the Midget that featured a hand operated gantry and this design fitted in well with the company's plans for modernising the delivery operation.*

Bottom Right: *The Midget was substantially redesigned by the versatile Mr. Needs along with assistance from Bill Hardy at the Peterborough garage, and manufactured (under license) in the UK by The Primrose Engineering Company. Known as the 'SELFSTAK' it consisted of a lightweight gantry able to travel along the body side channels. Attached to the gantry was a power-driven hoist and grab that could lift and move the units of 350 bricks and lower them to the offside of the lorry. The bricks themselves were loaded onto the lorries in blocks of 1,050 bricks by forklift truck and these large blocks easily subdivided into 350 block units. If necessary, when the bricks were lowered to the ground from lorries like Foden HMJ 920V, they could be further sub-divided into units or slices of 70 bricks, which could then be moved around the building site on special barrows.*

Above: *With a few modifications over the last 25 years the Selfstak remained lighter, simpler and less expensive to operate than conventional lorry cranes. This is not to say that the company ignored the improvements in crane technology in this period. A number of new vehicles like Foden KBH 959V were fitted with hydraulic cranes during the 1980s to assist with the more difficult deliveries. These cranes had the ability to swing anywhere within a 360° radius and so enable bricks to be placed just where they were needed and thus reduce the chance of damage to the facing bricks. In addition to the blocks of bricks being banded, there is now the facility for the blocks to be shrink-wrapped to reduce the chance of water damage to the bricks before delivery to zero. The company policy is now that all new vehicles will be fitted with hydraulic cranes, but continual developments are always being evaluated as the company strives to deliver its products in the quickest and safest manner possible.*

The growth of the road vehicle contrasted with the fairly rapid decline in the number of bricks carried by rail during the 1960s. In an attempt to reduce damage in transit British Railways introduced the PALBRICK wagon, which was designed specifically for the movement of bricks on pallets. These wagons featured screw adjusters to restrain the load, but the continuing incidence of damage combined with the decline in the number of goods stations able to handle bricks reduced the volumes transported. Having said that, in 1964 for instance the railway moved a total of 446 million bricks from the Peterborough area.

Unfortunately it was the era of Dr. Beeching, who had a great dislike of single rail wagons being sent to obscure destinations and who did his level best to stop this traffic by the simple expedient of closing all the small and medium sized goods yards. A number of the company's depots at main railheads remained in operation including those at King's Cross (London), Sheffield, Hull, Leeds, Tyne/Tees, Manchester and Liverpool. As late as 1974 some 10% of London Brick production was still being moved by rail. At the railheads the bricks, which by this time were loaded on pallets on the rail wagons, would be delivered by tippers with bodies fitted with cill bars to restrain the pallet whilst the bricks tumbled out.

The next development was the 'Fletliner' trains that could be forklift-loaded with blocks of 1,050 bricks. In the first instance the blocks were not banded, relying instead on fibreglass panels to restrain the bricks, these panels could be laid flat to enable the Fletliner bodies to be stacked together for return to Stewartby. These trains ran to Liverpool and Manchester and could move up to 1.5 million bricks per week. One very positive benefit from these trains was the development of high quality load restraints. Sadly the often changing nature and location of delivery points, combined with the fixed terminals and a number of union disputes did not help! Coupled with this was the fact that the LBC had to pay for a fixed number of trains each year, regardless of the number actually used.

Since flexibility is paramount within a distribution network, the rail operation eventually foundered, though a revival under rail privatisation is always possible. There was also a short experiment with Roadheads, here bricks were delivered by road to a central point in bulk on maximum weight articulated units. The bricks would then be delivered by road to the customers in smaller loads. Again the problems of flexibility arose and this scheme also faltered, though there are still some 'outstations' that keep stocks of bricks in towns like Northallerton, Bridgewater and Falkirk.

Left: *With the Freightliner network of services to major cities, LBC altered their railhead delivery service and made use of skeletal flat wagons operating within the network. The flats were essentially 20' long iron and wood platforms, and also fitted an AEC Marshal Major (Selfstak) lorry chassis. The first train left Stewartby on 18th June 1973, and for a while the system was reasonably efficient as London Brick trucks were also located at the railheads to deliver the bricks to the building sites.*

WORKS VEHICLES

Scant attention has been paid so far to the 'behind the scenes fleet', yet the very first LBC mechanical vehicle was in fact destined for internal operations. We have made mention of the war surplus Chevrolet tippers; these were replaced by some Ford 500E tippers. In turn these were successively replaced by Ford Thames Traders and then by 'D' series Fords. and Leyland Boxers.

In addition to the tipper lorries, a number of light commercials (mostly vans) have always been used for a variety of jobs, including the movement of hot food between the central kitchens and works canteens. The LBC vans included Austin K8's (known as the three-way van due to the three doors used on the factory-built vans), Fordson E83W, Austin J4 and Ford Transits amongst others.

To discuss all the types of vehicles used internally within the works would be a physical impossibility, and even a generalisation would prove a major task in its own right, for such was the diversity of the support vehicle fleet. Accordingly, the following pictures may serve as a very basic introduction to some of the fascinating road vehicles that were used on or around the various LBC sites. This subject deserves much wider attention, and we would like to hear from readers with personal experience in this area.

Above: *The ex-WD Matadors were replaced by a brace of Highwayman tractors K503/4. Of these K503 was an articulated tractor, whilst K504 (seen here) was a drawbar tractor.*

Right: *This 1946 Muir Hill dumper is licensed to work on the public highway; given the registration number EMJ 194, it carries the LBC fleet number D38.*

Top Left: *This Austin 6 is seen at the Fletton First Aid Centre, it is one of two ambulances (MJ 8301-2) purchased new in 1935.*

Above: *This Fordson E83W van, OLU 901 was attached to the Estates Department, having been purchased second-hand from Transformers Ltd. of Watford.*

Centre Left: *This ex-WD Austin K2 was purchased in 1947 and registered JPP 243; despite its basic nature it lasted until 1961.*

Above: *In the service of the Stores Department is A14, a Bedford K type integral van EBM 753.*

Bottom Left: *This Bedford S Type fire engine was acquired from the South Shields Borough Brigade at the young age of just two years for use at the Stewartby Works in 1956.*

Mention has been made earlier of the huge demand for bricks after the war that was met initially by using the stockpiled supplies. The company then set about increasing the workforce to replace many of the POWs who had worked in the brickworks, and by the autumn of 1946 a night shift was introduced, only to be taken off almost immediately as the government reduced demand by 'playing' with the economy. It took $2^1/_2$ years for demand to return to the levels where the firm was willing to increase the workforce and then they were forced to look to Eastern Europe for the new workers.

With demand continually increasing and suitable local labour being in very short supply, the company went recruiting in Italy and large numbers moved from the sunny climes of the Mediterranean to the slightly less hospitable lowlands of the Home Counties.

The introduction of new recruits to an area already short of housing caused accommodation problems, so the LBC built hostels for the single men to live in and found houses for the families in the surrounding areas. This posed transport problems in an area where there were few suitable public transport facilities operating to the works. Having brought the men all the way from Italy it was no real problem for the company to get them to work, they simply started their own bus services!

Of course it was not quite that simple: routes had to be worked out, and licenses for the services had to be approved; but the biggest problem to overcome was the supply of suitable buses. At that time new buses were impossible to get and secondhand ones were in great demand. Fortunately the LBC was able to obtain several 'very well-used' buses with which to commence their services.

The company was able to specify that the majority of buses were to come fitted with engines built by AEC, but at that point standardisation ended! A total of 60 Regents, Regals, Maudslays and Bristols joined the fleet over a 25 year period. Just to confuse the issue some Regents arrived as double-deckers, whilst others carried single-deck bodies and at least one changed from double to single whilst in the company's employ. One very interesting batch of 1931 AEC Regent Is carried the first production double-deck bodies built at the United Automobile's Lowestoft Coach factory. Later in 1931, this factory passed into the operational orbit of the Eastern Counties Omnibus Company and the factory title changed to the Eastern Counties, later to become the famous Eastern Coach Works.

A similar Regent that had originated with the South Wales company was altered from double to single deck. However, the batch of single-deckers from Morley's Grey Coaches had all been rebuilt and rebodied prior to purchase. One of these buses was a rebuilt 'stop gap' AEC Reliance from 1929, which had used the old AEC chassis with a new 6-cylinder engine. Once the major operators had implemented their own fleet replacement plans, vehicles of a much better quality came on to the market. The LBC were then able to purchase better buses from BET companies like City of Oxford and Maidstone & District.

Above: *Whilst most chassis utilised AEC engines, the bodies varied considerably. and this AEC Regent III has a Park Royal low-bridge body. Formerly with the City of Oxford fleet, SFC 436 was obtained in 1965 and given the LBC fleet number C25.*

Below: *The LBC bus fleet usually looked presentable from the outside, and all the buses were maintained to PSV standards. Yet this arrival from South Wales (Brush-bodied AEC Regent I, WN 4762) clearly displays a sagging body and body strengthening straps. It entered the fleet in July 1947, and is really showing its 1931 build date in this 1949 view. However, it was soon re-bodied with the use of a secondhand single-deck body, but it did retain the original petrol engine for a little while longer, although a diesel unit was eventually fitted.*

Above: *Not all members of the LBC bus fleet were of direct AEC origin, here LKT 996 a Bristol L6A with an ECW body but fitted with an AEC engine leaves Stewartby Works in 1964, - passing an ex-army DUKW amphibian.* Photo Courtesy. W.P. Badgery

All the services were directly tied to the workers shift patterns and often two buses on almost identical routes would follow each other into the works, spaced just a few minutes apart. This was caused by the differing shift patterns of say the loaders and the blockers and the fact that neither group would be seen in the other's bus. Despite these buses being the only way for many employees to get to work they were seen as being cold, cramped, dirty and always breaking down! The services continued into the early 1970s, but the all-pervading influence of the private car spread its shadow across the viability of the operation. The company gradually reduced the operation, although some bus services still exist these have been supplied by contractors since 1971.

The buses worked to set routes and timetables as agreed with the Traffic Commissioners. In fact the LBC operated buses just like any other company except that the destinations were never in the town centres and as stated, the internal furnishings left a lot to be desired. Some of the principal routes in Bedfordshire and Buckinghamshire are detailed below, but there may have been others:

Bedford to Arlesey
Bedford to Stewartby (three different routes)
Cranfield to Stewartby
Flitwick to Stewartby
Newport Pagnell to Calvert
Buckingham to Calvert
Silverstone to Calvert
Aylesbury to Calvert
Bicester to Calvert
Leighton Buzzard to Calvert
Bletchley to Newton Longueville.

THE LONDON BRICK COMPANY BUS FLEET

No.	REG	MODEL	BODY	YEAR	PURCHASED FROM	BOUGHT	SOLD
C1	GH 8091	Regal I	LT?	1932	London Transport	9/46	5/56
C2	WN 4762	Regent I	Brush	1931	South Wales	7/47	11/57
C3	RT 7723	Regent I	United Auto	1931	Lowestoft	7/47	12/51
C4	RT 7726	Regent I	United Auto	1931	Lowestoft	7/47	12/51
C5	RT 7728	Regent I	United Auto	1931	Lowestoft	7/47	12/51
C6	RT 7729	Regent I	United Auto	1931	Lowestoft	7/47	12/51
C7	RT 7724	Regent I	United Auto	1931	Lowestoft	7/47	3/52
C8	RT 7725	Regent I	United Auto	1931	Lowestoft	7/47	12/51
C10	JO 8662	Regent I	Park Royal	1934	City of Oxford	10/48	12/51
C11	CP 9065	Regal I	Park Royal	1931	Halifax Corp.	10/48	12/51
C12	KR 5853	Regal I	?	1930	Morley's Greys	12/48	5/56
C13	GV 9810	Regal I	Alexander	1931	Morley's Greys	12/48	11/57
C14	HX 4332	Regal I	Alexander	1931	Morley's Greys	12/48	11/57
C15	GJ 8074	Regal I	Alexander	1931	Morley's Greys	12/48	5/56
C16	ACF 10	Reliance(4-cyl)	Alexander	1929	Morley's Greys	12/48	6/55
C17	UY 8232	Regal I	Alexander	1931	Morley's Greys	4/49	2/55
C18	GJ 2017	Regent I	Tilling	1930	London Transport	1/49	6/56
C19	EV 3600	Regent I	Ransomes	1930	Colchester Corp.	5/49	4/53
C20	EV 3601	Regent I	Ransomes	1930	Colchester Corp.	5/49	7/52
C21	VX 5551	Regent I	Ransomes	1930	Colchester Corp.	2/49	7/52
C22	VX 5552	Regent I	Ransomes	1930	Colchester Corp.	8/49	7/53
C23	VX 5553	Regent I	Ransomes	1930	Colchester Corp.	6/49	6/53
C24	OF 3990	Regent I	Brush	1930	Birmingham	11/51	5/56
C25	OV 4474	Regent I	Brush	1930	Birmingham	11/51	5/56
C26	OV 4462	Regent I	Brush	1930	Birmingham	11/51	4/58
Second Batch							
C3	GL 5073	Regal I	ECW	1937	Bath Tramways	4/52	1/56
C4	GL 5082	Regal I	ECW	1937	Bath Tramways	4/52	1/60
C5	CUS 801	Regent I	EEC	1939	Glasgow Corp	6/53	5/57
C7	CUS 809	Regent I	EEC	1939	Glasgow Corp	5/53	1/60
C8	EF 8108	Regal I	Duple	1947	Richardson, H'pool	2/55	9/60
C9	HTM 988	Regal I	Duple	1949	Richardson, H'pool	8/55	1/56
C10	CAG 665	Maudslay II	Whitson	1947	Law, Prestwick	3/55	1/60
C11	HAU 695	Regal I	Duple	1946	Skill's, Nott'm	9/55	8/61
C12	CKG 290	Regent I	NCB	1943	Western Welsh	12/56	11/65
C16	DUK 755	Regal I	Plaxton	1946	Everall, Wolveh'n	9/55	9/66
C17	FXT 207	Regent I	LT (2RT)	1939	London Transport	2/56	1/66
C19	GOF 581	Regal I	Wadham	1947	Direct, B'ham	3/56	2/60
C20	EJF 771	Regal I	Duple	1947	Straw, Leics	4/56	1/60
C21	LVX 468	Maudslay	Whitson	1947	?	6/56	6/63
C22	EF 8107	Regal I	Duple	1947	Richardson, H'pool	6/56	6/63
C23	BAG 516	Regal	Duple	1946	Dodds of Troon	7/56	12/61
C27	SY 7811	Regal I	Duple	1946	Stewart, Dalkeith	10/56	10/63
Third Batch							
C3	KGK 758	Regent III	Craven	1948	London Transport	4/57	1/71
C4	NJO 710	Regal III	Willowbrook	1948	City of Oxford	8/60	1/71
C5	HKL 814	Regal I	Beadle	1946	Maidstone & Dist	10/57	9/66
C6	HKL 825	Regal I	Beadle	1946	Maidstone & Dist	10/57	4/65
C7	LKT 997	Bristol L6A	ECW	1950	Maidstone & Dist	10/57	4/65
C8	FFW 194	Bristol L6A	ECW	1949	Lincolnshire RCC	7/62	6/69
C9	ACP 407	Regent I	Park Royal	1947	Halifax Corp.	4/59	6/69
C10	NJO 718	Regal III	Willowbrook	1948	City of Oxford	5/63	11/67
C11	NJO 720	Regal III	Willowbrook	1948	City of Oxford	5/63	4/70
C13	HKL 855	Bristol K6A	Weymann	1946	Maidstone & Dist	9/59	3/66
C14	FWN 80	Regal III	Willowbrook	1949	South Wales	11/59	8/68
C15	FCY 341	Regal III	Willowbrook	1949	South Wales	11/59	11/67
C18	NJO 712	Regal III	Willowbrook	1948	City of Oxford	2/60	4/70
C19	OJO 721	Regal III	Willowbrook	1949	City of Oxford	11/63	3/70
C21	MLL 821	Regal IV	Metro Cammell	1951	London Transport	2/65	8/70
C22	MLL 577	Regal III	Metro Cammell	1951	London Transport	2/65	8/70
C23	SFC 733	Regal IV	Willowbrook	1952	City of Oxford	5/65	3/71
C24	LKT 996	Bristol L6A	ECW	1950	Maidstone & Dist	11/61	4/69
C25	SFC 436	Regent III	Park Royal	1952	City of Oxford	10/65	3/71
C26	JWN 905	Regent III	Weymann	1952	South Wales	12/65	1/71
C27	SFC 610	Regal IV	Willowbrook	1952	City of Oxford	12/65	1/71
Fourth Batch							
C1	SFC 439	Regent III	Park Royal	1952	City of Oxford	1/66	3/71
C2	UWL 929	Regent III	Park Royal	1954	City of Oxford	7/66	3/71
C6	TWL 178	Regent III	Weymann	1953	City of Oxford	7/66	3/71
C12	UEV 833	Bristol KSW5	ECW	1952	Eastern National	5/71	9/71

Returning to the subject of vehicle purchases, the AEC Marshals were followed by large numbers of Volvo F86 6-wheelers and later the improved Volvo F7 model entered service in six and 8-wheeler form. Amongst the AEC Marshal's the first batch to be equipped with Selfstak equipment came fitted with a larger, but de-rated AV 760 engine, still allied to the Mercury style axles and chassis frame in a weight-saving exercise. Unfortunately the extra torque produced by the larger engine ensured that differentials had a very short life on these chassis.

There was also a regular purchase of Seddon Atkinson 300 series 6-wheelers of which the early ones arrived fitted with International Harvester's own engine. The chassis and running gear of these 300s proved relatively trouble-free, with the exception of a campaign change on the gearboxes early in the working life of these vehicles.

The Volvos offered a feature that was brand new to Great Britain and that was the lifting rear axle. Despite the British operators attachment to double drive rear axles, the Swedish manufacturers regularly proved that a single driven rear axle allied with a lifting rear axle could often give better adhesion than the double drive in inclement conditions.

In addition the single lift axle could be lifted off the road when running light to save tyre wear and it also helped reduce the turning circle. At first this lifting axle was a mixed blessing, since with the axle lifted the 'swept circle' of the rear of the vehicle was much greater than before and very good at removing fixed roadside furniture until the drivers got used to the long rear overhang.

With the purchase of Marston Valley in 1968 came a fleet of 212 vehicles including ERF and Commer 4-wheelers with some of the ERFs dating back to 1957. Leyland group models included LAD cabbed Comets, AEC Mercurys and Marshalls.

The acquisition of Redland brought several Bedfords (mainly 16-ton KMs) and Foden 8-wheelers into LBC colours. A large number of the Fodens dated back to 1961 and 1962 and had originated with the Eastwoods company. Redland also donated four Volvo F86 articulated units. From the orange painted Marston Valley fleet the best remembered vehicles were probably the AEC Mammoth Major Mark V 8-wheelers which boasted a sliding door to the cab. Whilst this might have seemed the epitome of forward-thinking at the time of delivery, it relied on the drivers to be of only average size to actually get into the cab.

In Conclusion

From very humble beginnings the London Brick grew to become the major producer of bricks in the country and the biggest brick company in the world. With the huge advantage that was given to the brick-makers by the use of a raw material that would almost cook itself, the company could offer British house builders a virtually limitless supply of bricks.

It was able to produce bricks almost at the drop of the proverbial hat, and in turn it played a significant part in both the industrial and the urban development of many towns and cities in Britain. It was able to satisfy the demand for building materials as and when they arose, and it also provided much-needed bricks in Britain's darkest hour. In the aftermath of World War II the company helped to meet the needs of post-war reconstruction, addressing both the shortages in labour and transport in an innovative manner.

London Brick always operated a highly reliable and successful fleet that always gave the clear impression that its management were aware of the advertising potential of a clean and tidy vehicles, and the need to give customers the best possible service. In the 1960s the London Brick Company were the first company to adopt a successful and integrated mechanical handling process for the production, stacking, loading and delivery of bricks both by road and rail.

The adventurous use of the empty clay pits for the acceptance of waste led to another chapter in the growth of the company, followed by the take-over by the Hanson Group. Then came the new corporate image introduced in 1999! These have all given their own aspects of development, allowing the company to remain an industry leader.

At the start of the 21st Century one of the world's oldest crafts (brick-making) is seen to be as important now as it was in the 19th Century. The Hanson Group will of course continue to play an integral part in its development and promotion as research continues into new uses for bricks manufactured from THE CLAY THAT BURNS .

Finally, the publishers and I would like to extend our grateful thanks to the following people who have helped with the content of the book and offered much advice and assistance.

Andrew Bayles	Robert Berry
Carol Bellwood	Robin Chambers
Doreen Cooke	Mike Davies
Alan Earnshaw	Graham Edge
Richard Grey	Richard Hillier
Robin Hannay	John (Rommel) Hardy
Bill Harris	Peter Harris
Derek Jacson	David Ireson
Bill Needs	Ian Nichols
Leonard Mascapelli	Don Neugent
Peter Ousby	Alice Quiggley
David Percival	Mira Popovic
Bob Puttock	Jack Rowe
D. Showler	David Townend
Allan Utterthwiate	Antonio Zacanzio

plus the management and various members of staff (past and present) at Hanson Brick, and to the Lord James Hanson.